Evapotranspiration: Remote Sensing

Edited by **Elizabeth Lamb**

New York

Published by Callisto Reference,
106 Park Avenue, Suite 200,
New York, NY 10016, USA
www.callistoreference.com

Evapotranspiration: Remote Sensing
Edited by Elizabeth Lamb

International Standard Book Number: 978-1-63239-333-3 (Hardback)

Printed in the United States of America.

Contents

Preface

This book is the end result of constructive efforts and intensive research done by experts in this field. The aim of this book is to enlighten the readers with recent information in this area of research. The information provided in this profound book would serve as a valuable reference to students and researchers in this field.

This book contains several well researched topics associated with various aspects and properties of evapotranspiration (ET) including remote-sensing based balancing determination and simulation of ET. These fields are the leading technologies that measure the extremely spatial ET from the planet's surface. The analyses explain technicalities of ET simulation from incompletely vegetated surfaces and stomatal conductance behavior of natural and agricultural ecosystems. A crucial review explains strategies used in hydrological structures and the applications explain ET arrangements in alpine catchments, under water shortage, under climate modification, and for grasslands and pastures. Suggested guiding principles for applying operational satellite-based energy balance models and overcoming common challenges have also been included.

At the end, I would like to thank all the authors for devoting their precious time and providing their valuable contribution to this book. I would also like to express my gratitude to my fellow colleagues who encouraged me throughout the process.

Editor

Evapotranspiration Estimation Based on the Complementary Relationships

Virginia Venturini[1], Carlos Krepper[1,2] and Leticia Rodriguez[1]
*[1]Centro de Estudios Hidro-Ambientales-Facultad de Ingeniería y
Ciencias Hídricas Universidad Nacional del Litoral
[2]Consejo Nacional de Investigaciones Científicas y Técnicas
Argentina*

1. Introduction

Many hydrologic modeling and agricultural management applications require accurate estimates of the actual evapotranspiration (ET), the relative evaporation (F) and the evaporative fraction (EF). In this chapter, we define ET as the actual amount of water that is removed from a surface due to the processes of evaporation-transpiration whilst the potential evapotranspiration (Epot) is any other evaporation concept. There are as many potential concepts as developed mathematical formulations. In this chapter, F represents the ratio between ET and Epot, as it was introduced by Granger & Gray (1989). Meanwhile, EF is the ratio of latent flux over available energy.

It is worthy to note that, in general, the available evapotranspiration concepts and models involve three sets of variables, i.e. available net radiation (Rn), atmospheric water vapor content or temperature and the surface humidity. Hence, different Epot formulations were derived with one or two of those sets of variables. For instance, Penman (1948) established an equation by using the Rn and the air water vapor pressure. Priestley & Taylor (1972) derived their formulations with only the available Rn.

In the last three decades, several models have been developed to estimate ET for a wide range of spatial and temporal scales provided by remote sensing data. The methods could be categorized as proposed by Courault et al. (2005).

Empirical and semi-empirical methods: These methods use site specific or semi-empirical relationships between two o more variables. The models proposed by Priestley & Taylor (1972), hereafter referred to as P-T, Jackson et al. (1977); Seguin et al. (1989); Granger & Gray (1989); Holwill & Stewart (1992); Carlson et al. (1995); Jiang & Islam (2001) and Rivas & Caselles (2004), lie within this category.

Residual methods: This type of models commonly calculates the energy budged, then ET is estimated as the residual of the energy balance. The following models are examples of residual methods: The Surface Energy Balance Algorithm for Land (SEBAL) (Bastiaanssen et al., 1998; Bastiaanssen, 2000), the Surface Energy Balance System (SEBS) model (Su, 2002) and the two-source model proposed by Norman et al. (1995), among others.

Indirect methods: These physically based methods involve Soil-Vegetation-Atmosphere Transfer (SVAT) models, presenting different levels of complexity often reflected in the number of parameters. For example, the ISBA (Interactions between Soil, Biosphere, and

Atmosphere) model by Noilhan & Planton (1989), developed to be included within large scale meteorological models, parameterizes the land surface processes. The ISBA Ags model (Calvet et al., 1998) improved the canopy stomatal conductance and CO_2 concentration with respect to the ISBA original model.

Among the first category (Empirical and semi-empirical methods), only few methodologies to calculate ET have taken advantage of the complementary relationship (CR).

It is worth mentioning that there are only two CR approaches known so far, one attributed to Bouchet (1963) and the other to Granger & Gray (1989). Even though various ET models derived from these two fundamental approaches are referenced to throughout the chapter, it is not the intention of the authors to review them in detail.

Bouchet (1963) proposed the first complementary model based on an experimental design. He postulated that, for a large homogeneous surface and in absence of advection of heat and moisture, regional ET could be estimated as a complementary function of Epot and the wet environment evapotranspiration (Ew) for a wide range of available energy. Ew is the ET of a surface with unlimited moisture. Thus, if Epot is defined as the evaporation that would occur over a saturated surface, while the energy and atmospheric conditions remain unchanged, it seems reasonable to anticipate that Epot would decrease as ET increases. The underlying argument is that ET incorporates humidity to the surface sub-layer reducing the possibility for the atmosphere to transport that humidity away from the surface. Bouchet´s idea that Epot and ET have this complementary relationship has been the subject of many studies and discussions, mainly due to its empirical background (Brutsaert & Parlange, 1998; Ramírez et al., 2005). Examples of successful models based on Bouchet's heuristic relationship include those developed by Brutsaert & Stricker (1979); Morton (1983) and Hobbins et al. (2001). These models have been widely applied to a broad range of surface and atmospheric conditions (Brutsaert & Parlange, 1998; Sugita et al., 2001; Kahler & Brutsaert, 2006; Ozdogan et al., 2006; Lhomme & Guilioni, 2006; Szilagyi, 2007; Szilagyi & Jozsa, 2008).

Granger (1989a) developed a physically based complementary relationship after a meticulous analysis of potential evaporation concepts. He remarked that "*Bouchet corrected the misconception that a larger potential evaporation necessarily signified a larger actual evaporation*". The author used the term "potential evaporation" for the Epot and Ew concepts, and clearly presented the complementary behavior of common potential evaporation theories. This author suggested that Ew is the value of the potential evaporation when the actual evaporation rate is equal to the potential rate. The use of two potential parameters, i.e. Epot and Ew, seems to generate a universal relationship, and therefore, universal ET models. Conversely, attempting to estimate ET from only one potential formulation may need site-specific calibration or auxiliary relationships (Granger, 1989b). In addition, the relative evaporation coefficient introduced by Granger & Gray (1989) enhances the complementary relationship with a dimensionless coefficient that yields a simpler complementary model.

The foundation of the complementary relationship is the basis for operational estimates of areal ET by Morton (1983), who formulated the Complementary Relationship Areal Evapotranspiration (CRAE) model. The reliability of the independent operational estimates of areal evapotranspiration was tested with comparable, long-term water budget estimates for 143 river basins in North America, Africa, Ireland, Australia and New Zealand.

A procedure to calculate ET requiring only common meteorological data was presented by Brutsaert & Stricker (1979). Their Advection-Aridity approach (AA) is based on a conceptual model involving the effect of the regional advection on potential evaporation and Bouchet's complementary model. Thus, the aridity of the region is deduced from the regional advection of the drying power of the air. The authors validated their model in a rural watershed finding a good agreement between estimated daily ET and ET obtained with the energy budget method.

Morton's CRAE model was tested by Granger & Gray (1990) for field-size land units under a specific land use, for short intervals of time such as 1 to 10 days. They examined the CRAE model with respect to the algorithms used to describe different terms and its applicability to reduced spatial and temporal scales. The assumption in CRAE that the vapor transfer coefficient is independent of wind speed may lead to appreciable errors in computing ET. Comparisons of ET estimates and measurements demonstrated that the assumptions that the soil heat flux and storage terms are negligible, lead to large overestimation by the model during periods of soil thaw.

Hobbins et al. (2001) and Hobbins & Ramírez (2001) evaluated the implementations of the complementary relationship hypothesis for regional evapotranspiration using CRAE and AA models. Both models were assessed against independent estimates of regional evapotranspiration derived from long-term, large-scale water balances for 120 minimally impacted basins in the conterminous United States. The results suggested that CRAE model overestimates annual evapotranspiration by 2.5% of mean annual precipitation, whereas the AA model underestimates annual evapotranspiration by 10.6% of mean annual precipitation. Generally, increasing humidity leads to decreasing absolute errors for both models. On the contrary, increasing aridity leads to increasing overestimation by the CRAE model and underestimation by the AA model, except at high aridity basins, where the AA model overestimates evapotranspiration.

Three evapotranspiration models using the complementary relationship approach for estimating areal ET were evaluated by Xu & Singh (2005). The tested models were the CRAE model, the AA model, and the model proposed by Granger & Gray (1989) (GG), using the concept of relative evaporation. The ET estimates were compared in three study regions representing a wide geographic and climatic diversity: the NOPEX region in Central Sweden (typifying a cool temperate humid region), the Baixi catchment in Eastern China (typifying a subtropical, humid region), and the Potamos tou Pyrgou River catchment in Northwestern Cyprus (typifying a semiarid to arid region). The calculation was made on a daily basis whilst comparisons were made on monthly and annual bases. The results showed that using the original parameter values, all three complementary relationship models worked reasonably well for the temperate humid region, while their predictive power decreased as soil moisture exerts increasing control over the region, i.e. increased aridity. In such regions, the parameters need to be calibrated.

Ramírez et al. (2005) provided direct observational evidence of the complementary relationship in regional evapotranspiration hypothesized by Bouchet in 1963. They used independent observations of ET and Epot at a wide range of spatial scales. This work is the first to assemble a data set of direct observations demonstrating the complementary relationship between regional ET and Epot. These results provided strong evidence for the complementary relationship hypothesis, raising its status above that of a mere conjecture.

A drawback among the aforementioned complementary ET models is the use of Penman or Penman-Monteith equation (Monteith & Unsworth, 1990) to estimate Epot. Specifically, the Morton's CRAE model (Morton, 1983) uses Penman equation to calculate Epot, and a modified P-T equation to approximate Ew. Brutsaert & Stricker (1979) developed their AA model using Penman for Epot and the P-T equilibrium evaporation to model Ew. At the time those models were developed, networks of meteorological stations constituted the main source of atmospheric data, while the surface temperature (Ts) or the soil temperature were available only at some locations around the World. The advent of satellite technology provided routinely observations of the surface temperature, but the source of atmospheric data was still ancillary. Thus, many of the current remote sensing approaches were developed to estimate ET with little amount of atmospheric data (Price, 1990; Jiang & Islam, 2001).

The recent introduction of the Atmospheric Profiles Product derived from Moderate Resolution Imaging Spectroradiometer (MODIS) sensors onboard of EOS-Terra and EOS-Aqua satellites meant a significant advance for the scientific community. The MODIS Atmospheric profile product provides atmospheric and dew point temperature profiles on a daily basis at 20 vertical atmospheric pressure levels and at 5x5km of spatial resolution (Menzel et al., 2002). When combined with readily available Ts maps obtained from different sensors, this new remote source of atmospheric data provides a new opportunity to revise the complementary relationship concepts that relate ET and Epot (Crago & Crowley, 2005; Ramírez et al., 2005).

A new method to derive spatially distributed EF and ET maps from remotely sensed data without using auxiliary relationships such as those relating a vegetation index (VI) with the land surface temperature (Ts) or site-specific relationships, was proposed by Venturini et al. (2008). Their method for computing ET is based on Granger's complementary relationship, the P-T equation and a new parameter introduced to calculate the relative evaporation (F=ET/Epot). The ratio F can be expressed in terms of Tu, which is the temperature of the surface if it is brought to saturation without changing the actual surface vapor pressure. The concept of Tu proposed by these authors is analogous to the dew point temperature (Td) definition.

Szilagyi & Jozsa (2008) presented a long term ET calculation using the AA model. In their work the authors presented a novel method to calculate the equilibrium temperature of Ew and P-T equation that yields better long-term ET estimates. The relationship between ET and Epot was studied at daily and monthly scales with data from 210 stations distributed all across the USA. They reported that only the original Rome wind function of Penman yields a truly symmetric CR between ET and Epot which makes Epot estimates true potential evaporation values. In this case, the long-term mean value of evaporation from the modified AA model becomes similar to CRAE model, especially in arid environments with possible strong convection. An R^2 of approximately 0.95 was obtained for the 210 stations and all wind functions used. Likewise, Szilagyi & Jozsa in (2009) investigated the environmental conditions required for the complementary ET and Epot relationship to occur. In their work, the coupled turbulent diffusion equations of heat and vapor transport were solved under specific atmospheric, energy and surface conditions. Their results showed that, under near-neutral atmospheric conditions and a constant energy term at the evaporating surface, the analytical solution across a moisture discontinuity of the surface yields a symmetrical complementary relationship assuming a smooth wet area.

Recently, Crago et al. (2010) presented a modified AA model in which the specific humidity at the minimum daily temperature is assumed equal to the daily average specific humidity. The authors also modified the drying power calculation in Penman equation using Monin-Obukhov theory (Monin & Obukhov, 1954). They found promising results with these modifications. Han et al. (2011) proposed and verified a new evaporation model based on the AA model and the Granger's CR model (Granger, 1989b). This newly proposed model transformed Granger´s and AA models into similar, dimensionless forms by normalizing the equations with Penman potential model. The evaporation ratio (i.e. the ratio of ET to Penman potential evaporation) was expressed as a function of dimensionless variables based on radiation and atmospheric conditions. From the validation with ground observations, the authors concluded that the new model is an enhanced Granger`s model, with better evaporation predictions. In addition, the model somewhat approximates the AA model under neither too-wet nor too-dry conditions. As the reader can conclude, the complementary approach is nowadays the subject of many ongoing researches.

2. A review of Bouchet's and Granger's models

Bouchet (1963) set an experiment over a large homogeneous surface without advective effects. Initially, the surface was saturated and evaporated at potential rate. With time, the region dried, but a small parcel was kept saturated (see Figure 1), evaporating at potential rate. The region and the parcel scales were such that the atmosphere could be considered stable. Bouchet described his experiment, dimension and scales as follows[1],

- The energy balance requires the prior definition of the limits of the system. To avoid taking into account the phenomena of accumulation and restoration of heat during the day and night phases, the assessment will cover a period of 24 hours.
- The system includes an ensemble of vegetation, soil, and a portion of the lower atmosphere. The sizes of these layers are such that the daily temperature variations are not significant.
- If this system is located in an area which, for any reason, does not have the same climatic characteristics, there will be exchanges of energy throughout the side "walls" of the system, that need to be analyzed (advection free area).
- Lateral exchanges by conduction in the soil are negligible. The lateral exchanges in the atmosphere due to the homogenization of the air masses will be named as "oasis effect". Given the heterogeneity from one point to another, the lateral exchanges of energies, or the "oasis effect", rule the natural conditions.
- The oasis effect phenomenon can be schematically represented as shown in Figure 1. If in a flat, homogeneous area (brown line in Figure 1), a discontinuity appears, i.e. a change in soil specific heat, moisture or natural vegetation cover, etc. (green line in Figure 1), then a disturbed area is developed in the direction of airflow (gray filled area in Figure 1) where environmental factors are modified from the general climate because of the discontinuity.
- The perturbation raises less in height than in width. It always presents a "flat lens" shape in which the thickness is small compared to the horizontal dimensions.

As mentioned, initially the surface was saturated and evaporated at its potential rate, i.e. at the so-called reference evapotranspiration (or Ew). In this initial condition, Epot = Ew = ET. When ET is lower than Ew due to limited water availability, a certain excess of energy would become available. This remaining energy not used for evaporation may, in tern, warm the lower layer of the atmosphere. The resulting increase in air temperature due to the heating, and the decrease in humidity caused by the reduction of ET, would lead to a new value of Epot larger than Ew by the amount of energy left over.

---

[1] The following text was translated by the authors of this chapter from Bouchet's original paper (in French).

General Climate of the External System

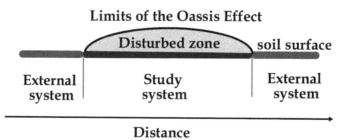

Fig. 1. Reproduction of Bouchet´s schematic representation of the Oasis Effect experiment.

Thus, Bouchet's complementary relationship was obtained from the balance of these evaporation rates,

$$ET + Epot = 2Ew$$

(1)

Bouchet postulated that in such a system, under a constant energy input and away from sharp discontinuities, there exists a complementary feedback mechanism between ET and Epot, that causes changes in each to be complementary, that is, a positive change in ET causes a negative change in Epot (Ozdogan et al., 2006), as sketched in Figure 2. Later, Morton (1969) utilized Bouchet's experiment to derive the potential evaporation as a manifestation of regional evapotranspiration, i.e. the evapotranspiration of an area so large that the heat and water vapor transfer from the surface controls the evaporative capacity of the lower atmosphere.

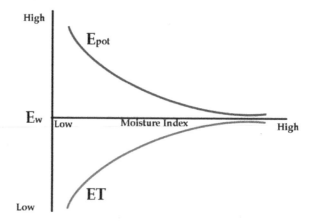

Fig. 2. Sketch of Bouchet´s complementary ET and Epot relationship

The hypothesis asserts that when ET falls below Ew as a result of limited moisture availability, a large quantity of energy becomes available for sensible heat flux that warms and dries the atmospheric boundary layer thereby causing Epot to increase, and *vise versa*.

Equation (1) holds true if the energy budget remains unchanged and all the excess energy goes into sensible heat (Ramírez et al., 2005). It should be noted that Bouchet´s experimental system is the so-called advection-free-surface in P-T formulation.

This relationship assumes that as ET increases, Epot decreases by the same amount, i.e. δET = -δEpot, where the symbol δ means small variations. Bouchet's equation has been widely used in conjunction with Penman (1948) and Priestley-Taylor (1972) (Brutsaert & Stricker, 1979; Morton, 1983; Hobbins el al., 2001).

Granger (1989b) argued that the above relationship lacked a theoretical background, mainly due to Bouchet's symmetry assumption (δET=-δEpot). Nonetheless, the author recognized that Bouchet´s CR set the basis for the complementary behavior between two potential concepts of evaporation and ET. One of the benefits of using two potential evaporation concepts rather than a single one is that the resulting CR would be universal, without the need of tuning parameters from local data.

Granger (1989a) revised the diversity of potential evaporation concepts available at that moment and expertly established an inequity among them. The resulting comparison yielded that Penman (1948) and Priestley & Taylor (1972) concepts are Ew concepts, and that the true potential evaporation would be that proposed by van Bavel (1966). Thus, these parameterizations would result in the following inequity, Epot ≥ Ew ≥ ET, where Epot would be van Bavel´s concept, Ew could be obtained with either Penman or P-T, knowing that ET-Penman is larger than ET-Priestley-Taylor (Granger, 1989a). Hence, the author postulated that the above inequity comprises Bouchet´s equity (δET = -δEpot) but it is based on a new CR. Granger (1989b) then proposed the following CR formulation,

$$ET + Epot\frac{\gamma}{\Delta} = Ew\left(\frac{\Delta + \gamma}{\Delta}\right) \tag{2}$$

where γ is the psychrometric constant and Δ is the slope of the saturation vapor pressure (SVP) curve.

Equation (2) shows that for constant available energy and atmospheric conditions, $-\gamma/\Delta$ is equal to the ratio δET/δEpot. In addition, this CR is not symmetric with respect to Ew. It can be easily verified that equation (2) is equivalent to equation (1) when $\gamma=\Delta$. The condition that the slope of the SVP curve equals the psychrometric constant is only true when the temperature is near 6 °C (Granger, 1989b). This has been widely tested (Granger & Gray, 1989; Crago & Crowley, 2005; Crago et al., 2005; Xu & Singh, 2005; Venturini et al., 2008; Venturini et al., 2011).

3. Bouchet`s versus Granger`s complementary models

A review of the two complementary models widely used for ET calculations was presented. Both methods are not only conceptually different, but also differ in their derivations. Mathematically speaking, Bouchet's complementary relationship (equation 1) results a simplification of Granger's complementary equation (equation 2) for the case $\Delta=\gamma$. Equations (1) and (2) can also be written, respectively, as follows,

$$\frac{1}{2}ET + \frac{1}{2}Epot = Ew \tag{3}$$

$$\left(\frac{\Delta}{\Delta+\gamma}\right)ET+\left(\frac{\gamma}{\Delta+\gamma}\right)Epot = Ew \tag{4}$$

The re-written Bouchet´s complementary model, equation (3), clearly expresses Ew as the middle point between the ET and the Epot processes. In contrast, the re-written Granger´s complementary relationship, equation (4), shows how both, ET and Epot contribute to Ew with different coefficients, the coefficients varying with the slope of the SVP curve at the air temperature Ta, since γ is commonly assumed constant. For clarity, Table 1 summarizes all symbols and definitions used in this Chapter.

Recently, Ramírez et al., (2005) discussed Bouchet's coefficient "2" with monthly average ground measurements. In their application, Epot was calculated with the Penman-Monteith equation and Ew with the P-T model. They concluded that the appropriate coefficient should be slightly lower than 2.

Venturini et al. (2008) and Venturini et al. (2011) introduced the concept of the relative evaporation, F= ET/Epot, proposed earlier by Granger & Gray (1989), along with P-T equation in both CR models. Thus, Epot is replaced by ET/F and Ew is equated to P-T equation. Hence, replacing Epot in equation (3),

$$ET + \frac{ET}{F} = k\,Ew \tag{5}$$

where k is Bouchet´s coefficient, originally assumed k=2
Then, when Ew is replaced in (5) by the P-T equation, results

$$ET\left(1+\frac{1}{F}\right) = ka\ (Rn-G)\frac{\Delta}{\Delta+\gamma} \tag{6}$$

where α is the P-T's coefficient, and the rest of the variables are defined in Table 1. Finally, Bouchet's CR is obtained by rearranging the terms in equation (6),

$$ET = ka\left(\frac{F}{F+1}\right)\left(\frac{\Delta}{\Delta+\gamma}\right)(Rn-G) \tag{7}$$

Following the same procedure with equation (4), the equivalent equation for Granger´s CR model is,

$$ET = \alpha\left(\frac{F\Delta}{F\Delta+\gamma}\right)(Rn-G) \tag{8}$$

It should be noted that the underlying assumptions of equation (7) are the same as those behind equation (8), plus the condition that Δ is approximately equal to γ.

Both, equations (7) and (8), require calculating the F parameter, otherwise the equations would have only theoretical advantages and would not be operative models. Venturini et al. (2008) developed an equation for F that can be estimated using MODIS products. Their F method is briefly presented here.

Consider the relative evaporation expression proposed by Granger & Gray (1989),

$$\frac{ET}{Epot} = \frac{f_u\,(e_s-e_a)}{f_u(e_s^*-e_a)} \tag{9}$$

where f_u is a function of the wind speed and vegetation height, e_s is the surface actual water vapor pressure, e_a is the air actual water vapor pressure, e^*_s is the surface saturation water vapor pressure.

Symbol	Definition
α	Priestley & Taylor's coefficient. $\alpha = 1.26$
Δ [hPa/°C]	Slope of the saturation water vapor pressure curve
γ [hPa/°C]	Psychrometric constant
λE [W m⁻²]	Latent heat flux density
e_a [hPa]	Air actual water vapor pressure at Td
e^*_a [hPa]	Air saturation water vapor pressure at Ta
e_s [hPa]	Surface actual water vapor pressure at Tu
e^*_s [hPa]	Surface saturation water vapor pressure at Ts
Ew [W m⁻²]	Evapotranspiration of wet environment
Epot [W m⁻²]	Potential evapotranspiration
f_u	Wind function
F	Relative evaporation coefficient of Venturini et al. (2008)
G [W m⁻²]	Soil heat flux
H [W m⁻²]	Sensible heat flux
Q [W m⁻²]	Available energy, (Rn –G)
Rn [W m⁻²]	Net radiation at the surface
Ta [°K] or [°C]	Air temperature
Td [°K] or [°C]	Dew point temperature
Ts [°K] or [°C]	Surface temperature
Tu [°K] or [°C]	Surface temperature if the surface is brought to saturation without changing e_s

Table 1. Symbols and units

This form of the relative evaporation equation needs readily available meteorological data. A key difficulty in applying equation (9) lies on the estimation of (e_s-e_a), since there is no simple way to relate e_s to any readily available surface temperature. Thus, a new temperature should be defined. Many studies have used temperature as a surrogate for vapor pressure (Monteith & Unsworth, 1990; Nishida et al., 2003). Although the relationship between vapor pressure and temperature is not linear, it is commonly linearized for small temperature differences. Hence, e_s and e_s^* should be related to soil+vegetation at a temperature that would account for water vapor pressure. Figure 3 shows the relationship between e_s, e^*_s and e_a and their corresponding temperatures; where e_u^* is the SVP at an unknown surface temperature Tu.

An analogy to the dew point temperature concept (Td) suggests that Tu would be the temperature of the surface if the surface is brought to saturation without changing the surface actual water vapor pressure. Accordingly, Tu must be lower than Ts if the surface is not saturated and close to Ts if the surface is saturated. Consequently, e_s could be derived from the temperature Tu. Although Tu may not possibly be observed in the same way as Td, it can be derived, for instance, from the slope of the exponential SVP curve as a function of Ts and Td. This calculation is further discussed later in this chapter.

Assuming that the surface saturation vapor pressure at Tu would be the actual soil vapor pressure and that the SVP can be linearized, $(e_s - e_a)$ can be approximated by $\Delta_1(Tu-Td)$ and $(e^*_s - e_a)$ by $\Delta_2(Ts-Td)$, respectively. Figure 3 shows a schematic of these concepts.

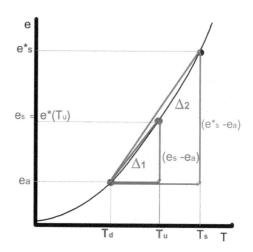

Fig. 3. Schematic of the linearized saturation vapor pressure curve and the relationship between $(e_s - e_a)$ and $\Delta_1(Tu-Td)$, and $(e^*_s - e_a)$ and $\Delta_2(Ts-Td)$.

Therefore, ET/Epot (see equation 9) can be rewritten as follows,

$$F = \frac{ET}{Epot} = \frac{(Tu-Td)}{(Ts-Td)}\left(\frac{\Delta_1}{\Delta_2}\right) \qquad (10)$$

The wind function, f_u, depends on the vegetation height and the wind speed, but it is independent of surface moisture. In other words, it is reasonable to expect that the wind function will affect ET and Epot in a similar fashion (Granger, 1989b), so its effect on ET and Epot cancels out. The slopes of the SVP curve, Δ_1 and Δ_2, can be computed from the SVP first derivative at Td and Ts without adding further complexity to this method. However, Δ_1 and Δ_2 will be assumed approximately equal from now on, as they will be estimated as the first derivative of the SVP at Ta.

The relationship between Ts and Tu can be examined throughout the definition of Tu, which represents the saturation temperature of the surface. For a saturated surface, Tu is expected to be very close or equal to Ts. In contrast, for a dry surface, Ts would be much larger than Tu. Since Epot is larger than or equal to ET, F ranges from 0 to 1. For a dry surface, with Ts >> Tu, (Ts-Td) would be larger than (Tu-Td) and ET/Epot would tend to 0. In the case of a saturated surface with e_s close to e_s^* and Ts close to Tu, (Ts-Td) would be similar to (Tu-Td) and ET/Epot would tend to 1.

The calculation of Tu proposed by Venturini et al. (2008) is presented in the next section, where results from MODIS data are shown. However, it is emphasized that the definition of Tu is not linked to any data source; therefore it can be estimated with different approaches.

4. Complementary models application using remotely sensed data

In order to show the potential of the complementary relationships, equations (7) and (8) were applied to the Southern Great Plains of the USA region and the results compared and analyzed.

4.1 Study area

The Southern Great Plains (SGP) region in the United States of America extends over the State of Oklahoma and southern parts of Kansas. The area broadens in longitude from 95.3° W to 99.5° W and in latitude from 34.5° N to 38.5° N (Figure 4). This region was the first field measurement site established by the Atmospheric Radiation Measurement (ARM) Program. At present, the ARM program has three experimental sites. Scientists from all over the World are using the information obtained from these sites to improve the performance of atmospheric general circulation models used for climate change research. The SGP was chosen as the first ARM field measurement site for several reasons, among them, its relatively homogeneous geography, easy accessibility, wide variability of climate cloud types, surface flux properties, and large seasonal variations in temperature and specific humidity (http://www.arm.gov/sites/sgp).

Most of this region is characterized by irregular plains. Altitudes range from approximately 500 m to 90 m, increasing gradually from East to West. In southwestern Oklahoma, the highest Wichita Mountains rise as much as 800 m above the surrounding landscape (Heilman & Brittin, 1989; Venturini et al., 2008). The climate is semiarid-subtropical. Although the maximum rainfall occurs in summer, high temperatures make summer relatively dry. Average annual temperatures range from 14°C to 18°C. Winters are cold and dry, and summers are warm to hot. The frost-free season stretches from 185 to 230 days. Precipitation ranges from 490 to 740 mm, with most of it falling as rain.

Grass is the dominant prairie vegetation. Most of it is moderately tall and usually grows in bunches. The most prevalent type of grassland is the bluestem prairie (*Andropogon gerardii and Andropogon hallii*), along with many species of wildflowers and legumes. In many places where grazing and fire are controlled, deciduous forest is encroaching on the prairies.

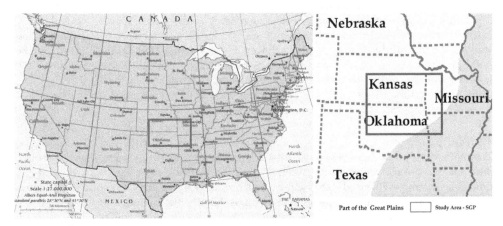

Fig. 4. Study area map

Due to generally favorable conditions of climate and soil, most of the area is cultivated, and little of the original vegetation remains intact. Oak savanna occurs along the eastern border of the region and along some of the major river valleys.

4.2 Ground data availability

The latent heat data was obtained from the ARM program Web site (http://www.arm.gov). The ARM instruments and measurement applications are well established and have been used for validation purposes in many studies (Halldin & Lindroth, 1992; Fritschen & Simpson, 1989). The site and name, elevation, geographic coordinates (latitude and longitude) and surface cover of the stations used in this work are shown in Table 2.

Site	Elevation (m a.m.s.l.)	Lat/Lon	Vegetation Type
Ashton, Kansas E-9	386	37.133 N/97.266 W	Pasture
Coldwater, Kansas E-8	664	37.333 N/99.309 W	Rangeland (grazed)
Cordell, Oklahoma: E-22	465	35.354 N/98.977 W	Rangeland (grazed)
Cyril, Oklahoma: E-24	409	34.883 N/98.205 W	Wheat (gypsum hill)
Earlsboro, Oklahoma: E-27	300	35.269 N/96.740 W	Pasture
Elk Falls, Kansas E-7	283	37.383 N/96.180 W	Pasture
El Reno, Oklahoma: E-19	421	35.557 N/98.017 W	Pasture (ungrazed)
Hillsboro, Kansas E-2	447	38.305 N/97.301 W	Grass
Lamont, Oklahoma: E-13	318	36.605 N/97.485 W	Pasture and wheat
Meeker, Oklahoma: E-20	309	35.564 N/96.988 W	Pasture
Morris, Oklahoma: E-18	217	35.687 N/95.856 W	Pasture (ungrazed)
Pawhuska, Oklahoma: E-12	331	36.841 N/96.427 W	Native prairie
Plevna, Kansas E-4	513	37.953 N/98.329 W	Rangeland (ungrazed)
Ringwood, Oklahoma: E-15	418	36.431 N/98.284 W	Pasture

Table 2. Site name and station name, elevation, latitude, longitude and surface type

The first instrumentation installation to the SGP site took place in 1992, with data processing capabilities incrementally added in the succeeding years. This region has relatively extensive and well-distributed coverage of surface fluxes and meteorological observation stations. In this study, Energy Balance Bowen Ratio stations (EBBR), maintained by the ARM program were used for the validation of surface fluxes. The EBBR system produces 30 minute estimates of the vertical fluxes of sensible and latent heat at the local points. The EBBR fluxes estimates are calculated from observations of net radiation, soil surface heat flux, the vertical gradients of temperature and relative humidity.

4.3 MODIS products

The method proposed here was physically derived from universal relationships. Moreover, data sources do not represent a limitation for the applicability of equations (6) and (8), nonetheless remotely sensed data such as that provided by MODIS scientific team would empower the potential applications of the methods. Hence, the equations applicability using MODIS products was explored. The sensor's bands specifications can be obtained from http://modis.gsfc.nasa.gov/about/specifications.php.

Daytime images for seven days in year 2003 with at least 80% of the study area free of clouds were selected. Table 3 summarizes the images information including date, day of the year, satellite overpass time and image quality.

Geolocation is the process by which scientists specify where a specific radiance signal was detected on the Earth's surface. The MODIS geolocation dataset, called MOD03, includes eight Earth location data fields, e.g. geodetic latitude and longitude, height above the Earth ellipsoid, satellite zenith angle, satellite azimuth, range to the satellite, solar zenith angle, and solar azimuth. Similarly Earth location algorithms are widely used in modeling and geometrically correct image data from the Land Remote Sensing Satellite (Landsat) Multispectral Scanner (MSS), Landsat Thematic Mapper (TM), System pour l'Observation de la Terre (SPOT), and Advanced Very High Resolution Radiometer (AVHRR) missions.

Date in 2003	Day of the Year (DOY)	Overpass time (UTC)	Image Quality (% clouds)
March 23rd	82	17:05	18
March 31st	90	17:55	15
April 1st	91	17:00	18
September 6th	249	17:10	6
September 19th	262	16:40	23
October 12th	285	16:45	9
October 19th	292	16:50	6

Table 3. Date, Day of the Year, overpass time and image quality of the seven study days.

MOD11 is the Land Surface Temperature (LST), and emissivity product, providing per-pixel temperature and emissivity values. Average temperatures are extracted in Kelvin with a day/night LST algorithm applied to a pair of MODIS daytime and nighttime observations. This method yields 1 K accuracy for materials with known emissivities, and the view angle information is included in each LST product. The LST algorithms use other MODIS data as input, including geolocation, radiance, cloud masking, atmospheric temperature, water vapor, snow, and land cover. These products are validated, meaning that product uncertainties are well defined over a range of representative conditions. The theories behind this product can be found in Wan (1999), available at http://modis.gsfc.nasa.gov/data/atbd/atbd_mod11.pdf.

In particular, MODIS Atmospheric Profile product consists on several parameters: total ozone burden, atmospheric stability, temperature and moisture profiles, and atmospheric water vapor. All of these parameters are produced day and night at 5×5 km pixel resolution. There are two MODIS Atmosphere Profile data product files: MOD07_L2, containing data collected from the Terra platform and MYD07_L2 collecting data from Aqua platform. The MODIS temperature and moisture profiles are defined at 20 vertical levels. A simultaneous direct physical solution to the infrared radiative-transfer equation in a cloudless sky is used. The profiles are also utilized to correct for atmospheric effects for some of the MODIS products (e.g., sea-surface temperature and LST, ocean aerosol properties, etc) as well as to characterize the atmosphere for global greenhouse studies. Temperature and moisture profile retrieval algorithms are adapted from the International TIROS Operational Vertical Sounder (TOVS) Processing Package (ITPP), taking into account MODIS' lack of stratospheric channels and far higher horizontal resolution. The profile retrieval algorithm requires calibrated, navigated, and co-registered 1-km field of the view (FOV) radiances from MODIS channels 20, 22-25, 27-29, and 30-36. The atmospheric water vapor is most directly obtained by integrating the moisture profile through the atmospheric column. Data validation was conducted by comparing results from the Aqua platform with *in situ* data (Menzel et al., 2002). In the present study, air temperature and dew point temperature at 1000 hPa level are used to calculate the vapor pressure deficit. Also the temperatures are assumed to be homogenous over the 5x5 km grid.

5. Results

In this section, the results are divided in two parts. The results of variables and parameters needed to apply the CR models are presented in first place, followed by a comparison of results between equations (7) and (8).

5.1 Variables calculation

In order to apply Bouchet's and Granger's CR, Rn, G and F for each pixel of every image of the study area must be computed. The other parameters, Δ and γ, can be assumed constant for the entire region. Alternatively, they can be estimated with spatially distributed information of Ta over the region. The constants α and k are assumed equal to 1.26 and 2, respectively.

The Rn maps were estimated with the methodology published by Bisht et al. (2005), which provides a spatially consistent and distributed Rn map over a large domain for clear sky days. With this method, Rn can be evaluated in terms of its components of downward and upward short wave radiation fluxes, and downward and upward long wave radiation fluxes. Several MODIS data products are utilized to estimate every component. Details of these calculations for the study days presented in this work can be found in Bisht et al. (2005), from where we took the Rn maps.

Soil heat fluxes G were calculated according to Moran et al. (1989) with the daily Normalized Difference Vegetation Index (NDVI) maps (Kogan et al., 2003), calculated with MOD021KM products. The equations used are

$$G = 0.583 \text{ Rn } e^{(2.13*NDVI)} \qquad \text{for NDVI} > 0 \qquad (11)$$

$$G = 0.583 \text{ Rn} \qquad \text{for NDVI} \leq 0 \qquad (12)$$

The slope of the SVP curve, Δ, was calculated at Ta using Buck's equation *(Buck, 1981)* and the MODIS Ta product.

In order to determine F, a methodology to estimate Tu is needed. By definition, different types of soils and water content would render different Tu values. Here, it is proposed to estimate the variable Tu from the SVP curve. It can be assumed that e_s is larger or equal to e_a and lower or equal to e^*_s, thus Tu must lie between Ts and Td.

The first derivative of the SVP curve at Ts and at Td represents the slope of the curve between those points. It can also be computed from the linearized SVP curve between the intervals [Tu,Ts] and [Td,Tu], which are symbolized as Δ_1 and Δ_2, respectively. Thus, an expression for Tu is derived from a simple system of two equations with two unknowns, as follows,

$$T_u = \frac{\left(e_s^* - e_a\right) - \Delta_1 Ts + \Delta_2 Td}{\Delta_2 - \Delta_1} \qquad (13)$$

There are many published SVP equations that can be used to obtain the derivative of e as function of the temperature. Here, Buck's formulation (Buck, 1981) was chosen for its simple form (equation 14),

$$e = 6.1121 \exp\left(\frac{17.502 \ T}{240.97 + T}\right) \tag{14}$$

where "e" is water vapor pressure [hPa] and T is temperature [°C]. Thus, the first derivative of equation 14 is computed at Td and Ts to estimate Δ_1 and Δ_2 in equation (13).

$$\frac{de}{dT} = \left[\frac{4217.45694}{(240.97 + T)^2}\right] *6.1121 \exp\left(\frac{17.502 \ T}{240.97 + T}\right) \tag{15}$$

The estimation of Tu could be improved by introducing another surface variable, such as soil moisture or any other surface variable that accounts for the surface wetness. However, in order to demonstrate the strength of the CR models, the Tu calculation is kept simple, with minimum data requirements. It is recognized, however, that this calculation simplifies the physical process and may introduce errors and uncertainties to the F ratio.

Figure 5 shows Rn maps obtained for April 1st, 2003 as an example of what can be expected in terms of spatial resolution with Bisht et al. methodology. Figure 6 displays Tu map for the same date obtained with the MOD07 spatial resolution (5x5 km).

5.2 Comparison of the CR models

The results obtained from equations (7) and (8) are compared to demonstrate the strength of the complementary relationship. The contrasted results were computed assuming k=2, α=1.26, γ=0.67 hPa/C, Δ was obtained with Ta maps, estimating F as proposed in Venturini et al. (2008). The resulting ET estimates are shown in Table 4, where average root mean square errors (RMSEs) and biases are about 25 Wm-2, indicating that equation (7), obtained with Bouchet`s complementary model, would lead to larger ET estimates. However, only the "ground truth" would tell which equation is more precise. In this case, the ground truth is considered to be the ground measurements of ET described in section 4.2. Then, observed ET values were compared with the results obtained using equations (7) and (8), (see Figure 7). The overall RMSE is about 52.29 and the bias (Observed-Bouchet) is –37.90 Wm-2. For Granger`s CR, the overall RMSE and bias (Observed-Granger) are 33.89 and -10.96 Wm-2 respectively, with an R^2 of about 0.79.

	RMSE	BIAS (Bouchet-Granger)	R^2
DOY82	5.42	0.91	0.990
DOY90	7.38	0.86	0.993
DOY91	13.70	13.01	0.983
DOY 249	31.74	31.56	0.995
DOY 262	25.51	25.33	0.991
DOY 285	26.79	26.40	0.990
DOY 292	28.24	28.11	0.999

Table 4. ET(Wm-2) comparison between Bouchet´s and Granger´s CR.

From Table 4 it can be concluded that Bouchet's simplification results in larger ET estimates, with biases up to approximately 32 Wm⁻², than those obtained with Granger's CR. From Figure 7 it can be seen that Bouchet's CR overestimates ground observations as well.

Ramírez et al. (2005) derived the value of Bouchet's k parameter from ground data. The authors presented evidences of the complementary relationship from independent measurements of ET and Epot. Then, k values were calculated for different hypothesis. These authors reported a mean k of about 2.21 and a k variance equal to 0.07 using uncorrected pan evaporation data as a surrogate of Epot.

In this chapter, equations (7) and (8) are equated and k calculated for instantaneous ET values. Thus,

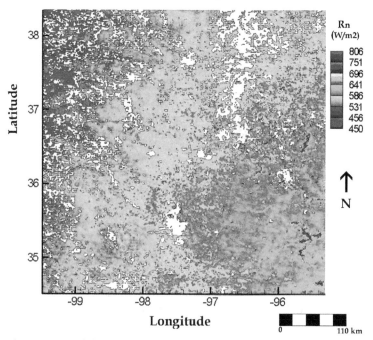

Fig. 5. Net radiation map of the SGP for April 1st, 2003

$$\frac{F\Delta}{F\Delta + \gamma} = \frac{kF\Delta}{(F+1)(\Delta + \gamma)} \tag{16}$$

$$k = \frac{(F+1)(\Delta + \gamma)}{F\Delta + \gamma} \tag{17}$$

Bouchet's coefficient k was calculated for each pixel in every day. The overall mean k value is 2.341, with an overall minimum of 1.784 and a maximum of 2.710, standard deviations varying from 0.025 to 0.078. These results are close to those reported by Ramírez et al. (2005).

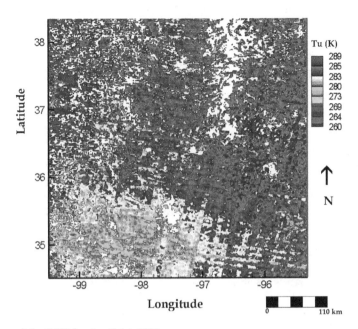

Fig. 6. Tu map of the SGP for April 1st, 2003

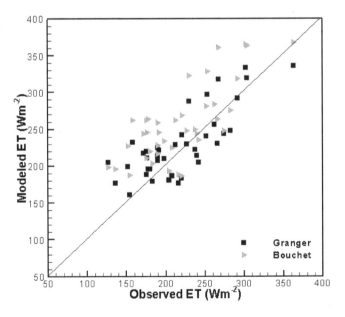

Fig. 7. Comparison between Bouchet's and Granger`s complementary models against ground measurements

Both complementary models yield similar ET estimates, however Granger's model lead to more accurate results than Bouchet's method. The slope of the SVP curve at the air temperature sets a k value slightly different from 2.

6. Spatial and temporal scales considerations

The complementary theory assumes a surface without advection influences and so does the regional evapotranspiration concept (Penman, 1948; Priestly & Taylor, 1972; Brutsaert & Stricker 1979). In fact, in his original work, Bouchet (1963) described five scales implicated in the oasis effect (see Table 5). Therefore for each scale of heterogeneity (s), we can define the oasis effects that give the lateral energy exchange of Q_1, Q_2, Q_3, Q_4, Q_5. In the development of his theory he assumed that only Q_3 is variable with ET while Q_4 and Q_5 are not affected by changes of ET and Epot associated with water availability. For the other two scales, s_1 and s_2, Q_1 and Q_2 are not involved in the complementary relationship. Bouchet's experiment established an energy balance over 24 hours, avoiding taking into account the phenomena of accumulation and restoration of heat during the day and night phases. These particular assumptions left smaller time and space scales out of the CR, therefore a review of the scales of applicability of the CR might be interesting.

The "evaporation paradox" mentioned by Brutsaert & Parlange (1998) refers to the seemingly opposing trends observed between pan evaporation and actual evaporation. The authors suggested that the paradox is solved in the CR framework.

The usefulness of the CR for understanding global scale in climate studies have been analized by Brutsaert & Parlange (1998), Szilagyi (2001) and Hobbins et al. (2001), among others. Szilagy & Josza, (2009), coupled Bouchet's CR with a long-term water-energy balance based on considerations of the precipitation time series and the soil water balance. The authors show that important ecosystem characteristics, such as the maximum soil water storage, can be derived from this "long-term" application of the CR. The scales shown in Table 5 seem to be compatible with those used in the aforementioned works. Nonetheless, the applicability of the CR at small scales is not evident from Bouchet's publication.

Crago & Crowley (2005) evaluated the complementary relationship at relatively small temporal scales (10 to 30 min) using data from meteorological stations in different grassland sites. The authors demonstrated that the CR holds true also at small scales. Kahler and Brutsaert (2006) used properly scaled data of daily ET and daily pan evaporation observed at two experimental sites to demonstrate the validity of the CR. The CR at daily scales was confirmed by this research. The authors argue that for unscaled daily data of pan evaporation the CR may not be noticeable.

Scale (symbol in the text)	Timescales	Spatial Scales	Effects of oasis corresponding
Molecular - s_1	10^9 second	few hundred meters	Q_1
Turbulent - s_2	1 second to some minutes	few hundred meters	Q_2
Convection and related movements - s_3	10 minutes to a few hours	few kilometers	Q_3
Cyclonic - s_4	3 to 4 hours	1000 a 2000 kilometers	Q_4
Global - s_5	10 to 30 hours	5000 to 10000 km	Q_5

Table 5. Translation of Table 1 published by Bouchet in 1963

In a more practical way, the method proposed by Venturini et al. (2008) corrects the ET from a saturated surface with the local surface-atmosphere conditions at the pixel scale. The absence of regional assumptions makes the method applicable to a wide range of spatial scales even though the background of their method is Granger´s CR. Venturini´s method has been applied with instantaneous data, i.e. remotely sensed data with MODIS. The comparison between observed and estimated ET values yields errors of about 15% of observed instantaneous ET(Venturini et al., 2011).

7. References

Bastiaanssen, W.G.M., Menenti, M.A, Feddes, R.A. & Hollslag, A.A.M. (1998). A remote sensing surface energy balance algorithm for land (SEBAL) 1. Formulation. *Journal of Hydrology*, 212, 13, pp. 198-212, ISSN 0022-1694.

Bastiaanssen, W.G.M. (2000). SEBAL-based sensible and latent heat fluxes in the irrigated Gediz Basin, Turkey. *Journal of Hydrology*, 229, pp. 87-100, ISSN 0022-1694.

Bisht, G., Venturini, V., Jiang, L. & Islam, S. (2005). Estimation of Net Radiation using MODIS (Moderate Resolution Imaging Spectroradiometer) Terra Data for clear sky days. *Remote Sensing of Environment*, 97, pp. 52-67, ISSN 0034-4257.

Bouchet, R.J. (1963). Evapotranspiration rèelle et potentielle, signification climatique. *International Association of Scientific Hydrology*, 62, pp. 134-142, ISSN 0262-6667.

Brutsaert, W., & Stricker, H. (1979). An advection-aridity approach to estimate actual regional evapotranspiration. *Water Resources Research*, 15,2, pp. 443–450, ISSN 0043-1397.

Brutsaert W., & Parlange M.B. (1998) Hydrologic cycle explains the evaporation paradox. *Nature*, 396, pp. 30, ISSN 0028-0836.

Buck, A.L. (1981). New equations for computing vapor pressure and enhancement factor. *Journal of Applied Meteorology*, 20, pp. 1527-1532, ISSN 0894-8763.

Calvet, J.C., Noilhan , J. & Besseoulin, P. (1998). Retrieving the root zone soil moisture from surface soil moisture or temperature estimates: A feasibility study on field measurements. *Journal of Applied Meteorology*, 37, pp. 371-386, ISSN 0894-8763.

Carlson, T.N., Gillies, R.R., & Schmugge, T. J. (1995). An interpretation of methodologies for indirect measurement of soil water content. *Agricultural and Forest Meteorology*, 77, pp. 191-205, ISSN 0168-1923.

Courault, D., Seguin, B. & Olioso, A. (2005). Review to estimate Evapotranspiration from remote sensing data: Some examples from the simplified relationship to the use of mesoscale atmospheric models. *Irrigation and Drainage Systems*, 19, pp. 223-249, ISSN 0168-6291.

Crago, R., & Crowley, R. (2005). Complementary relationship for near-instantaneous evaporation. *Journal of Hydrology*, 300, pp. 199-211, ISSN 0022-1694.

Crago, R., Hervol, N., & Crowley, R. (2005). A complementary evaporation approach to the scalar roughness length. *Water Resources Research*, 41, W06017, ISSN 0043-1397.

Crago, R.D., Qualls R.J., & Feller M. (2010) A calibrated advection-aridity evaporation model requiring no humidity data. *Water Resources Research*, 46, W09519, doi:10.1029/2009WR008497, (September, 2010), ISSN. 0043-1397.

Fritschen, L., & Simpson, J. R. (1989). Surface energy and radiation balance systems: General description and improvements. *Journal of Applied Meteorology* 28, 680-689. ISSN 0894-8763.

Granger, R.J. (1989a). An examination of the concept of potential evaporation. *Journal of Hydrology*, 111, pp. 9-19, ISSN 0022-1694.

Granger, R.J., & Gray, D.M. (1989). Evaporation from natural nonsaturated surfaces. *Journal of Hydrology*, 111, pp. 21-29, ISSN 0022-1694.

Granger, R.J. (1989b). A complementary relationship approach for evaporation from nonsaturated surfaces. *Journal of Hydrology*, 111, pp. 31-38, ISSN 0022-1694

Granger, R.J., & Gray, D.M. (1990). Examination of Morton's CRAE model for estimating daily evaporation from field-sized areas. *Journal of Hydrology*, 120, pp. 309-325, ISSN 0022-1694.

Halldin, S & Lindroth, A. (1992). Errors in net radiometry: Comparison and evaluation of six radiometer designs. *Journal of Atmospheric Oceanic Technology*, 9, 762-783, ISSN 0739-0572.

Han, S., Hu, H., Yang, D., & Tian, F. (2011). A complementary relationship evaporation model referring to the Granger model and the advection–aridity model. *Hydrological Processes*, 25, 8, doi:10.1002/hyp.7960, ISSN 0885-6087.

Heilman, J.L. & Brittin, C. L. (1989). Fetch requirements for Bowen ratio measurements of latent and sensible heat fluxes. *Agricultural and Forest Meteorology*, 44, 261-273, ISSN 0168-1923.

Hobbins, M.T., & Ramírez, J.A. (2001). The complementary relationship in estimation of regional evapotranspiration: An enhanced advection-aridity model, *Water Resources Research*, 37,5, pp. 1389-1403, ISSN 0043-1397.

Hobbins, M.T., Ramírez, J.A., Brown T.C. & Classens L.H.J.M. (2001). The complementary relationship in estimation of regional evapotranspiration: The complementary relationship areal evapotranspiration and advection-aridity models, *Water Resources Research*, 37,5, pp. 1367-1487, ISSN 0043-1397.

Holwill, C.J., & Stewart, J.B. (1992). Spatial variability of evaporation derived from Aircraft and ground-based data. *Journal of Geophysical Research*, 97, D17, pp. 19061-19089, ISSN 0148-0227

Jackson, R.D., Reginato, R.J., & Idso, S.B. (1977). Wheat canopy temperature: A practical tool for evaluating water requirements. *Water Resources Research*, 13, pp. 651-656, ISSN 0043-1397

Jiang, L. & Islam, S. (2001). Estimation of surface evaporation map over southern Great Plains using remote sensing data. *Water Resources Research*, 37(2), 329-340. ISSN 0043-1397.

Kahler, D. M. & Brutsaert, W. (2006). Complementary relatinship between daily evaporation in the environment and pan evaporation. *Water Resources Research*, 42, W05413, doi:10.1029/2005WR004541, ISSN 0043-1397.

Kogan, F., Gitelson, A., Zakarin, E., Spivak, L. & Lebed, L., (2003). AVHRR-Based Spectral Vegetation Index for Quantitative Assessment of Vegetation State and Productivity: Calibration and Validation. *Photogrammetric Engineering & Remote Sensing*, 69, 8, (August 2003) pp. 899-906, ISSN 0099-1112.

Lhomme J.P., & Guilione L. (2006). Comments on some articles about the complementary relationship, Discussion. *Journal of Hydrology*, 323, pp. 1-3, ISSN 0022-1694.

Menzel, W.P., Seemann, S.W., Li, J., & Gumley, L.E. (2002). *MODIS Atmospheric Profile Retrieval Algorithm Theoretical Basis Document, Version 6. Reference Number: ATBD-MOD-07.* NASA. http://modis.gsfc.nasa.gov/data/atbd/atbd_mod07.pdf.

Monin, A.S. & Obukhov, A.M., (1954). Osnovnye zakonomernosti turbulentnogo peremesivanija v prizemnom sloe atmosfery. *Trudy Geofizicheskogo Instituta Akademiya Nauk SSSR*, 24, 151, pp. 163-187.

Monteith, J.L. & Unsworth, M. (1990). *Principles of Environmental Physics* (2nd edition), Butterworth-Heinemann, ISBN: 071312931X, Burlington-MA- USA.

Moran, M.S., Jackson, R.D., Raymond, L.H, Gay, L.W. & Slater, P.N. (1989). Mapping surface energy balance components by combining LandSat thematic mapper and ground-based meteorological data, *Remote Sensing of Environment*, 30, pp.77-87, ISSN 0034-4257.

Morton, F. I. (1969). Potential evaporation as manifestation of regional evaporation. *Water Resources Research*, 5, pp. 1244-1255, ISSN 0043-1397.

Morton, F.I. (1983). Operational estimates of areal evapotranspiration and their significance to the science and practice of hydrology. *Journal of Hydrology*, 66, 1-76, ISSN 0022-1694.

Nishida, K., Nemani, R.R., Running, S.W. & Glassy, J.M. (2003). An operational remote sensing algorithm of land evaporation. *Journal of Geophysical Research*, 108, D9, 4270, doi:10.1029/2002JD002062, ISSN 0148-0227.

Noilhan, J. & Planton, S. (1989). GCM gridscale evaporation from mesoscale modelling. *Journal of Climate*, 8, pp. 206-223, ISSN 0894-8755

Norman, J.M., Kustas, W.P. & Humes, K.S. (1995). Sources approach for Estimating soil and vegetation energy fluxes in observations of directional radiometric surface temperature. *Agricultural Forest and Meteorology*, 77, pp. 263-293, ISSN 0168-1923.

Ozdogan M., Salvucci G.D., & Anderson B.T. (2006). Examination of the Bouchet-Morton complementary relationship using a mesoscale climate model and observations under a progressive irrigation scenario, *Journal of Hydrometeorology*, 7, pp. 235-251, ISSN 1525-755X

Penman, H.L. (1948). Natural evaporation from open water, bare soil and grass. *Proceedings of the Royal Society of London, Series A*, 193,1032,(April, 1948), pp. 120-145, ISSN 1471-2946.

Price, J.C. (1990). Using spatial context in satellite data to infer regional scale evapotranspiration. *IEEE Transactions on Geoscience and Remote Sensing*, 28, 5, pp. 940-948, ISSN 0196-2892

Priestley, C.H.B. & Taylor, R.J. (1972). On the Assessment of Surface Heat Flux and Evaporation Using Large-Scale Parameters. *Monthly Weather Review*, 100, pp. 81–92, ISSN 0027-0644.

Ramírez, J.A., Hobbins, M.T. & Brown T. (2005). Observational evidence of the complementary relationship in regional evaporation lends strong support for Bouchet's hypothesis. *Geophysical Research Letters*, 32, L15401, doi:10.1029/2005GL023549, ISSN 0094-8276.

Rivas, R. & Caselles, V. (2004). A simplified equation to estimate spatial reference evaporation from remote sensing-based surface temperature and local meteorological data. *Remote Sensing of Environment*, 83, pp. 68-76, ISSN 0034-4257.

Seguin, B., Assad, E., Fretaud, J.P., Imbernom, J.P., Kerr, Y., & Lagouarde, J.P. (1989). Use of meteorological satellite for rainfall and evaporation monitoring. *International Journal of Remote Sensing*, 10, pp. 1001-1017, ISSN 0143-1161.

Su, B. (2002). The surface energy balance system (SEBS) for estimation of turbulent heat fluxes. *Hydrology and Earth System Sciences*, 6, pp. 85-99, ISSN 1027-5606.

Sugita, M., Usui, J., Tamagawa, I. & Kaihotsu, I. (2001).Complementary relationship with a convective boundary layer to estimate regional evaporation. *Water Resources Research*, 37,2, pp. 353-365, ISSN 0043-1397.

Szilagyi J., (2001). On Bouchet's complementary hypothesis. *Journal of Hydrology*, 246, pp. 155-158, ISSN 0022-1694.

Szilagyi J. (2007). On the inherent asymmetric nature of the complementary relationship of evaporation, *Geophysical Research Letters*, 34, L02405, ISSN 0094-8276.

Szilagyi J., & Jozsa J. (2008). New findings about the complementary relationship based evaporation estimation methods. *Journal of Hydrology*, 354, pp. 171– 186, ISSN 0022-1694.

Szilagyi J., & Jozsa J. (2009). Analytical solution of the coupled 2-D turbulent heat and vapor transport equations and the complementary relationship of evaporation . *Journal of Hydrology*, 372, pp. 61–67, ISSN 0022-1694.

van Bavel, C.H.M. (1966). Potential evaporation: The combination concept and its experimental verification. *Water Resources Research* , 2, pp. 455-467, ISSN 0043-1397.

Venturini, V., Islam, S., & Rodríguez, L., (2008). Estimation of evaporative fraction and evapotranspiration from MODIS products using a complementary based model. *Remote Sensing of Environment*. 112, pp. 132-141, ISSN 0034-4257.

Venturini, V., Rodriguez L., & Bisht G. (2011). A comparison among different modified Priestley and Taylor´s equation to calculate actual evapotranspiration". *International Journal of Remote Sensing*, In Press, ISSN 0143-1161

Xu, C.Y., & Singh, V.P. (2005). Evaluation of three complementary relationship evapotranspiration models by water balance approach to estimate actual regional evapotranspiration in different climatic regions. *Journal of Hydrology*, 308, pp. 105-121, ISSN 0022-1694.

Wan, Z. (1999). *MODIS Land-Surface Temperature Algorithm Basis Document (LST ATBD)*, version 3.3, NASA, www.icess.ucsb.edu/ modis/atbd-mod-11.pdf.

2

Evapotranspiration Estimation Using Soil Water Balance, Weather and Crop Data

Ketema Tilahun Zeleke and Leonard John Wade
School of Agricultural and Wine Sciences, EH Graham Centre
for Agricultural Innovation, Charles Sturt University
Australia

1. Introduction

The rise in water demand for agriculture, industry, domestic, and environmental needs requires sagacious use of this limited resource. Since agriculture (mainly irrigation) is the major user of water, improving agricultural water management is essential. Efficient agricultural water management requires reliable estimation of crop water requirement (evapotranspiration). Evapotranspiration (ET) is the transfer of water from the soil surface (evaporation) and plants (transpiration) to the atmosphere. ET is a critical component of water balance at plot, field, farm, catchment, basin or global level. From an agricultural point of view, ET determines the amount of water to be applied through artificial means (irrigation). Reliable estimation of ET is important in that it determines the size of canals, pumps, and dams. The use of the terms 'reference evapotranspiration', 'potential evapotranspiration', 'crop evapotranspiration', 'actual evapotranspiration' in this chapter is based on FAO-56 (FAO Irrigation and Drainage publication No 56) (Allen et al., 1998).

There are different methods of determining evapotranspiration: direct measurement, indirect methods from weather data and soil water balance. These methods can be generally classified as empirical methods (eg. Thornthwaite, 1948; Blaney and Criddle, 1950) and physical based methods (eg. Penman, 1948; Montheith, 1981 and FAO Penman Montheith (Allen et al., (1998)). They vary in terms of data requirement and accuracy. At present, the FAO Penman Montheith approach is considered as a standard method for ET estimation in agriculture (Allen et al., 1998). A case study from a semiarid region of Australia will be used to demonstrate ET estimation for a canola (*Brassica napus* L.) crop using soil water balance and crop coefficient approaches. Daily rainfall data, soil moisture measurement data using neutron probe, and AquaCrop (Steduto et al., 2009) -estimated deep percolation below the crop root zone will be used to determine actual evapotranspiration of the crop using soil water balance. Reference evapotranspiration ET_o will be determined using FAO ET_o *calculator* (Raes, 2009). Crop canopy cover measured using a handheld *GreenSeeker^{TM}* and expressed as normalized difference vegetation index (NDVI) will be used to interpret evolution of evapotranspiration during the growing season (life cycle) of the canola crop.

2. Field experiment

2.1 Description of study area and field experiment

The study area is in Wagga Wagga, New South Wales (Australia). Wagga Wagga, referred to as 'the capital of Riverina', is located in the Riverina region of NSW. The Riverina extends from the foot hills of the Great Dividing Range in the east to the flat and dry inland plains in the west. Agriculture in the Riverina is significantly diversified with dry land farming of winter cereals and irrigation in Murrumbidgee and Colleambally irrigation areas. It has a Mediterranean type climate with a mixed farming system of winter cereal crops, summer crops, and pastures grazing lands. In addition to the major grain crops of rice, canola, wheat, and maize, the area also produces a quarter of NSW fruit and vegetable production (RDA, 2011). The Riverina region is characterized by the semiarid climate, with hot summers and cool winters (Stern et al., 2000). Seasonal temperature varies little across the region. More consistent rainfall occurs in winter months. Mean annual temperature is 15-18°C. January is the hottest month of the year while July is the coolest. Mean annual rainfall varies from 238 mm in the west to 617 mm in the east. Long term and 2010 mean monthly rainfall, reference evapotranspiration, and temperature are presented in Fig. 1. Rainfall in 2010 was much higher than the long term average while evapotranspiration in 2010 was lower than the long term average.

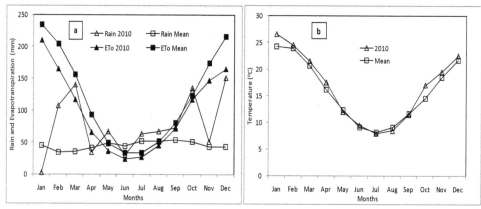

Fig. 1. (a) Rain and reference evapotranspiration ET_0 (long term average and in 2010) (b) Monthly average temperature (long term average and in 2010) at Wagga Wagga, NSW (Australia).

A field experiment was carried out during the growing season of 2010 at canola field experimental site of Wagga Wagga Agricultural Research Institute located at Wagga Wagga (35°03′N; 147°21′E; 235 m asl), NSW (Australia). There was enough rainfall (930 mm) in contrast to long term average of 522 mm in 2010 to provode ideal growing conditions. A popular variety of canola (Hyola50) was sown on 30 April 2010. The experiment was conducted on a 24 m x 24 m area. There were 24 plots, 12 experimental plots and 12 buffer plots. The plots were 6 m long with 1 meter buffer on either end. Plot width was 1.8 m with a 0.5 m walking strip between plots for data collection.

About a month before the experimental season, neutron probe access tubes were installed to a depth of 1.5 m for soil moisture measurement. Two access tubes were installed at 2 m from either end of the plot and 2 m from each other. Soil moisture content was measured at 15, 30, 45, 60, 90, and 120 cm depths every two weeks. The probe was calibrated using gravimetric soil moisture measurements done when access tubes were installed on site.

2.2 Weather data
Daily weather data (rainfall, minimum and maximum temperature, solar radiation, relative humidity, and wind speed) were collected from the meteorological station of the Wagga Wagga Agricultural Institute located adjacent to the experimental site. Out of the total annual rainfall of 930 mm, the amount or proportion (in percentage) during the canola growing season (May to November) was 514 mm (53%) while the long term average was 333 mm (64% of the long term average of 522 mm). Monthly average maximum and minimum temperature was 26°C and 3°C respectively. Reference evapotranspiration ET_0 was calculated using the procedure described in the FAO Irrigation and Drainage Paper 56 (Allen et al., 1998) with the help of the program FAO ET_0 Calculator (Raes, 2009).

2.3 Soil hydraulic characteristics
A 1.5m x 1.5m x 1.5m soil trench was dug for soil texture, field capacity (θ_{FC}), and wilting point (θ_{WP}) determination. Soil samples were retrieved from 0-30, 30-60, 60-90, and 90-120 cm depths for soil texture, θ_{FC}, and θ_{WP} determination using standard laboratory procedures hydrometer and pressure plate apparatus apparatus.

2.4 Crop parameters
The following crop phenological stages were recorded during the growing season: planting date, 90% emergence, beginning and end of flowering, senescence and maturity. The canopy cover was measured using GreenSeekerTM, an Optical Sensor Unit (NTech Industries, Inc., USA). GreenSeekerTM, is a handheld tool that determines Normalized Difference Vegetative Index (NDVI), is an integrated optical sensing and application system that measures green crop canopy cover.

3. Soil water balance method

Rain or irrigation reaching a unit area of soil surface, may infiltrate into the soil, or leave the area as surface runoff. The infiltrated water may (a) evaporate directly from the soil surface, (b) taken up by plants for growth or transpiration, (c) drain downward beyond the root zone as deep percolation, or (d) accumulate within the root zone. The water balance method is based on the conservation of mass which states that change in soil water content ΔS of a root zone of a crop is equal to the difference between the amount of water added to the root zone, Q_i, and the amount of water withdrawn from it, Q_o (Hillel, 1998) in a given time interval expressed as in Eq. (1).

$$\Delta S = Q_i - Q_o \tag{1}$$

Eq. (1) can be used to determine evapotranspiration of a given crop as follows

$$ET = P + I + U - R - D - \Delta S \tag{2}$$

where ΔS = change in root zone soil moisture storage, P = Precipitation, I = Irrigation, U = upward capillary rise into the root zone, R = Runoff, D = Deep percolation beyond the root zone, ET = evapotranspiration. All quantities are expressed as volume of water per unit land area (depth units).

In order to use Eq. (2) to determine evapotranspiration (ET), other parameters must be measured or estimated. It is relatively easy to measure the amount of water added to the field by rain and irrigation. In agricultural fields, the amount of runoff is generally small so is often considered negligible. When the groundwater table is deep, capillary rise U is negligible. The most difficult parameter to measure is deep percolation D. If soil water potential and moisture content are monitored, D can be estimated using Darcy's Principle. In this study, deep percolation estimated using AquaCrop (Raes et al., 2009), was adopted. Runoff R was also estimated using AquaCrop following USDA curve number approach (Hawkins et al., 1985). The change in soil water storage ΔS is measured using specialized instruments such as neutron probe and time-domain reflectrometer.

4. Crop coefficient method

4.1 Introduction

The crop coefficient approach relates evapotranspiration from a reference crop surface (ET_o) to evapotranspiration from a given crop (ET_c) through a coefficient. Estimation of crop water requirement from weather and crop data is a simpler and cost effective method compared to other methods such as soil water balance method. In this method, potential evapotranspiration of a crop is presumed to be determined by the evaporative demand of the atmosphere and crop characteristics. Evaporative demand of the air is determined as the evapotranspiration from a reference crop. The reference crop is a hypothetical crop (grass or alfalfa) with specific characteristics such as crop height of 0.12 m and albedo of 0.23 (Allen et al., 1998). Penman (1956) defined reference evapotranspiration as "the amount of water transpired in unit time by a shorter green crop, completely shading the ground, of uniform height and never short of water." It is a useful standard of reference for the comparison of different regions and of different measured evapotranspiration values within a given region. As such, ET_o is a climatic parameter expressing the evaporation power of the atmosphere independent of crop type, crop development and management practices (Allen et al., 1998). FAO Penman Montheith approach is considered as the standard method. In this method, reference evapotranspiration ET_o is estimated from weather data as given in Eq. (3).

$$ET_o = \frac{0.408\Delta\left(R_n - G\right) + \gamma \dfrac{900}{T + 273} u_2\left(e_s - e_a\right)}{\Delta + \gamma\left(1 + 0.34 u_2\right)} \tag{3}$$

where ET_o = reference evapotranspiration (mm/day); Rn = net radiation at the crop surface (MJ/m^2 day); G = soil heat flux density (MG/m^2 day); T = air temperature at 2 m height (°C); u_2 = wind speed at 2 m height (m/s); e_s= saturation vapor pressure (kPa); e_a = actual vapor pressure (kPa); e_s-e_a = saturation vapor pressure deficit (kPa); Δ = slope vapor pressure curve (kPa/°C); γ = psychrometric constant (kPa/°C).

Reference evapotranspiration ET_o can be calculated using a spreadsheet or computer programs which are designed for various level of data availability eg. *CROPWAT* (Smith, 1992) and *ET_o Calculator* (Raes, 2009). In this study, the latter program was used. It is important to make clear distinction between reference evapotranspiration ET_o and potential crop evapotranspiration ET_c. The latter is also called maximum crop evapotranspiration.

Evapotranspiration from a given crop grown and managed under standard conditions is called potential crop evapotranspiration ET_c. Standard condition is a disease-free, well-fertilized crops, grown in large fields, under optimum soil water conditions, and achieving full production under the given climatic conditions. ET_o depends evapotranspiration (ET_c) represents the climatic "demand" for water by a given crop. Potential crop depends primarily on the evaporative demand of the air.

4.2 Single crop coefficient method

The single crop coefficient (K_c) method is used to determine soil evaporation and transpiration lumped over a number of days or weeks. The single "time-averaged" K_c curve incorporates averaged transpiration and soil wetting effects into a single K_c factor. The FAO-56 publication divides the crop growth stages into four phenological stages. Initial stage is from planting to 10% ground cover. Development stage is from 10% groundcover to maximum cover. Midseason stage is from the beginning of full cover to the start of senescence. The late season stage is from the start of senescence to full senescence or harvest. The evolution of crop coefficients during these stages is tabulated in FAO-56 for a number of crops including canola. Three coefficients are given for the initial, midseason, and end of season stages as $K_{c\ ini}$, $K_{c\ mid}$, and $K_{c\ end}$ respectively. $K_{c\ ini}$ is assumed to be constant and relatively small (<0.4). The K_c begins to increase during the crop development stage and reaches a maximum value $K_{c\ mid}$ which is relatively constant for most growing and cultural conditions. During the late season period, as leaves begin to age and senesce, the K_c begins to decrease until it reaches a lower value at the end of the growing period equal to $K_{c\ end}$. The K_c during the development is estimated using linear interpolation between $K_{c\ ini}$ and $K_{c\ mid}$. Similarly, K_c during the late season stage is determined using linear interpolation between $K_{c\ mid}$ and $K_{c\ end}$. The value of $K_{c\ ini}$ and $K_{c\ end}$ can vary considerably on a daily basis, depending on the frequency of wetting by irrigation and rainfall. The single crop coefficient method can be used for irrigation planning and design. It is accurate enough for systems with large interval such as surface and set sprinkler irrigation. It is also used for catchment level hydrologic water balance studies (Allen et al., 1998).

In the single crop coefficient method, potential crop evapotranspiration ET_c is estimated from a single crop coefficient (K_c) and reference evapotranspirations ET_o as in Eq. (4).

$$ET_c = ET_o K_c \qquad (4)$$

Eq. (4) gives the potential (maximum) evapotranspiration of the crop when the soil moisture is not limiting. Since localized K_c values are not always available in many parts of the world, the values of K_c as suggested by FAO (Allen et al., 1998) are being widely used to estimate evapotranspiration.

When rainfall amount and irrigation are not sufficient to keep the soil moisture high enough, the soil moisture content in the root zone is reduced to levels too low to sustain the potential crop evapotranspiration ET_c. This results in an evapotranspiration less than the potential, and the plants are said to be under water stress. This evapotranspiration is called actual evapotranspiration (ET_a). In general, the actual evapotranspiration ET_a from various crops will not be equal to the potential value ET_c. Actual evapotranspiration ET_a is generally a fraction of ET_c depending on soil moisture availability. Actual evapotranspiration ET_a from a well-watered crop might generally approach ET_c during the active growing stage, but may fall below during the early growth stage, prior to full canopy coverage, and again toward the end of the growing season as the matured plant starts to dry out (Hillel, 1997). The actual evapotranspiration ET_a is calculated by combining the effects of K_c and soil water stress coefficient (K_s) as shown in Eq. (5).

$$ET_a = ET_oK_cK_s \tag{5}$$

The stress reduction coefficient K_s [0-1] reduces K_c when the average soil water content of the root zone is not high enough to sustain full crop transpiration. The stress coefficient K_s is determined by the amount of moisture the crop depleted from the rootzone of a crop. The amount of water depleted from the rootzone is expressed by root zone depletion D_r, i.e. water storage relative to field capacity. Stress is presumed to initiate when D_r exceeds the readily available water (RAW), Fig. 2. When more than RAW is extracted from the rootzone $(D_r > RAW)$, K_s is expressed (Allen et al., 1998) as

$$K_s = \frac{TAW - D_r}{TAW - RAW} = \frac{TAW - D_r}{(1 - p)TAW} \tag{6}$$

Where TAW = total plant available soil water in the root zone (mm), and p = fraction of TAW that a crop can extract from the root zone without suffering water stress. When $D_r \leq$ RAW, $K_s = 1$ indicating no water stress. The total available water in the root zone (TAW, mm) is estimated as the difference between the water content at the field capacity and wilting point

$$TAW = 1000\left(\theta_{FC} - \theta_{WP}\right)Z_r \tag{7}$$

Where Z_r = effective rooting depth (m); θ_{FC} is soil moisture content at field capacity (m³ m⁻³); θ_{WP} is soil moisture content at permanent wilting point (m³ m⁻³).

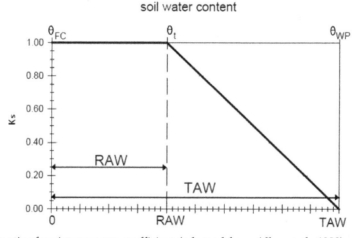

Fig. 2. Schematic of moisture stress coefficient (adapted from Allen et al., 1998).

Readily available water (RAW) is the amount of water which the crop can extract without experiencing stress. It is expressed as

$$RAW = pTAW \tag{8}$$

Soil moisture depletion fraction (p) is the fraction of soil water in the root zone that can be depleted before stress occurred. It varies from crop to crop and also varies at different growth stages of a given crop. Shallow rooted and sensitive crops such as vegetables have low p value while deep rooted and stress tolerant crops have a higher p value.

Canola crop coefficient values given in FAO 56 (Allen et al., 1998) are $K_{c\ ini}$ = 0.35, $K_{c\ mid}$ = 1.0-1.15, $K_{c\ end}$ = 0.35. These values represent K_c for a sub humid climate with RHmin = 45% and wind speed of 2 m/s. To take account for impacts of differences in aerodynamic roughness between crops and the grass reference with changing climate, the $K_{c\ mid}$ and $K_{c\ end}$ values larger than 0.45 must be adjusted using the following equation:

$$K_c = K_{c\ (tab)} + \left[0.04(u_2 - 2) - 0.004(RH_{min} - 45) \right] \left(\frac{h}{3} \right)^{0.3} \tag{9}$$

Where K_c (tab) is the value of K_c taken from Table 12 of Allen et al. (1998); h is the mean plant height during the mid or late season stage (m); RHmin the mean value for daily minimum relative humidity during the mid or late season growth stages (%) for 20%≤RHmins≤ 80%; u_2 is the mean value for daily wind speed at 2 m during the mid season or late season stages (m/s) for 1m/s ≤ u_2 ≤ 6 m/s. In this study, $K_{c\ ini}$ = 0.35, $K_{c\ mid}$ = 1.10, and $K_{c\ end}$ = 0.35 were used. Accordingly, $K_{c\ mid}$ value was adjusted to 1.08 for RHmin = 48%, u_2 = 1.91 m/s, and plant height of 1.0 m during this stage. Since $K_{c\ end}$ was less than 0.4, it was not necessary to adjust it. Once the K_{cb} values for the initial stage, mid season stage, and end-of-season stage were determined, K_{cb} values for development and late season stages were determined using linear interpolation.

4.3 Dual crop coefficient method

The single coefficient method does not separate evaporation and transpiration components of evapotranspiration. The dual crop coefficient approach calculates the actual increase in K_c for each day as a function of plant development and the wetness of the soil surface. It is best for high frequency irrigation such as microirrigation, centre pivots, and linear move systems (Suleiman et al., 2007). The effects of crop transpiration and soil evaporation are determined separately using two coefficients: the basal crop coefficient (K_{cb}) to describe plant transpiration and the soil water evaporation coefficient (K_e) to describe evaporation from the soil surface, Eq (10). AquaCrop determines crop transpiration (T_r) and soil evaporation (E) by multiplying ETo with their specific coefficients K_{cb} and K_e (Eq. 11) (Steduto et al., 2009).

$$K_c = K_{cb} + K_e, \text{ and} \tag{10}$$

$$ET_c = (K_{cb} + K_e) ET_o \tag{11}$$

The range of K_{cb} and K_e is [0-1.4]. When soil moisture is limiting, K_{cb} is multiplied by a coefficient K_s which is equal to 1 when D_r≤RAW and declines linearly to zero when all the available water in the rooting zone has been used. Evapotranspiration under such a condition is calculated using Eq. (12).

$$ET_a = (K_sK_{cb} + K_e) ET_o \tag{12}$$

Because the water stress coefficient impacts only crop transpiration, rather than evaporation from the soil, the application using Eq. (12) is generally more valid than is application using

Eq. (5) in the single crop coefficient approach. Allen et al. (1998) reported that in situations where evaporation from soil is not a large component of ET_c, use of Eq. (5) will provide reasonable results. The dual coefficient approach can be summarized into the following three steps: Calculate reference evapotranspiration (ET_0) from climatic data using Eq. (3), calculate individual crops potential evapotranspiration ET_c using Eq. (11), and when the soil moisture content is limited, K_{cb} coefficient is multiplied by stress factors K_s to calculate actual evapotranspiration ET_a using Eq. (12).

4.3.1 Basal crop coefficient

The basal crop coefficient K_{cb} is defined as the ratio of ET_c to ET_0 when the soil surface layer is dry but where the average soil water content of the rootzone is adequate to sustain full plant transpiration (Bonder et al., 2007). The dual crop coefficient approach uses daily time step and is readily adapted to spreadsheet program. Some models such as AquaCrop (Steduto et al., 2009) determine crop water productivity from the "productive" component of evapotranspiration i.e. transpiration. AquaCrop requires regression of daily values of biomass and crop transpiration to determine crop water productivity. Therefore, transpiration should be measured or estimated.

FAO-56 has tabulated K_{cb} values for a number of crops, including canola, at the initial, mid season, and end of season stages. Since localized K_{cb} values were not available for the study area, the values of K_{cb} suggested by FAO-56 (Allen et al., 1998) were used. For canola these value were $K_{cb\ ini}$ = 0.15, $K_{cb\ mid}$ = 0.95-1.10, and $K_{cb\ end}$ = 0.25. In this study, K_{cb} of 0.15, 1, and 0.25, respectively, for the initial, mid-season, and end of season stages were selected. The growing season of canola vary from 5 months to 7 months in Australia i.e. 150 -210 days depending on the planting date and the weather conditions (rainfall and temperature) during the season. Initial, development, mid-season, and late season stage lengths for canola grown during the 2010 winter season in Wagga Wagga (Australia) were 10, 64, 84, 48 days respectively.

The values for K_{cb} in the FAO-56 table represent values for a sub humid climate with RH_{min} = 45% and wind speed of 2 m/s. To take account for impacts of differences in aerodynamic roughness between crops and the grass reference, the $K_{cb\ mid}$ and $K_{cb\ end}$ values larger than 0.45 must be adjusted using the following equation:

$$K_{cb} = K_{cb\ (tab)} + \left[0.04(u_2 - 2) - 0.004(RH_{min} - 45)\right]\left(\frac{h}{3}\right)^3 \qquad (13)$$

Where K_{cb} (tab) is the value of $K_{cb\ mid}$ taken from Table 17 of Allen et al. (1998). The other parameters are as defined in Eq. (9). The K_{cb} values for the mid-season stage was adjusted using Eq. (13) to 0.98 for for RH_{min} = 48%, u_2 = 1.91 m/s, and plant height of 1.0 m. Once the K_{cb} values for the initial stage, mid season stage, and end-of-season stage were determined, K_{cb} values for development and late season stages were determined using linear interpolation.

The K_{cb} coefficient for any period (day) of the growing season can be derived by considering that during the initial and mid-season stages K_{cb} is constant and equal to the K_{cb} value of the growth stage under consideration. During the crop development and late season stage, K_{cb} varies linearly between the K_{cb} at the end of the initial stage ($K_{c\ ini}$) and the K_{cb} at the beginning of the midseason stage ($K_{cb\ mid}$). During the mid season stage K_{cb} is constant as $K_{cb\ mid}$. During late season stage, K_{cb} varies linearly between K_{cb} mid and K_{cb}

end. In the case of canola the end of season K_{cb} does not need adjustment since it is 0.25 which is less than 0.45.

4.3.2 Soil evaporation coefficient

Similar to K_{cb}, soil evaporation coefficient K_e needs to be calculated on a daily basis. K_e is a function of soil water characteristics, exposed and wetted soil fraction, and top layer soil water balance (Allen et al., 2005). In the initial stage of crop growth, the fraction of soil surface covered by the crop is small, and thus, soil evaporation losses are considerable. Following rain or irrigation, K_e can be as high as 1. When the soil surface is dry, K_e is small and even zero. K_e is determined using Eq. (14).

$$K_e = \min\{[K_r(Kc \max - K_{cb})],[f_{ew} Kc \max]\} \tag{14}$$

Where K_c max = maximum value of crop coefficient K_c following rain or irrigation; K_r = evaporation reduction coefficient which depends on the cumulative depth of water depleted; and f_{ew} = fraction of the soil that is both wetted and exposed to solar radiation. K_c max represents an upper limit on evaporation and transpiration from the cropped surface. K_c max ranges [1.05-1.30] (Allen et al., 2005). Its value is calculated for initial, development, mid-season, or late season using Eq. 15.

$$K_{c\ max} = \max\left(\left\{1.2 + \left[0.04(u_2 - 2) - 0.004(RH_{min} - 45)\right]\left(\frac{h}{3}\right)^{0.3}\right\},\{K_{cb} + 0.05\}\right) \tag{15}$$

Evaporation occurs predominantly from the exposed soil fraction. Hence, evaporation is restricted at any moment by the energy available at the exposed soil fraction, i.e. K_e cannot exceed f_{ew} x K_c max. The calculation of K_e consists in determining K_c max, K_r, and f_{ew}. K_c max for initial, development, midseason, and late season stages were calculated to be 1.196, 1.181, 1.187, and 1.195 respectively.

4.3.3 Evaporation reduction coefficient

The estimation of evaporation reduction coefficient K_r requires a daily water balance computation for the surface soil layer. Evaporation from exposed soil takes place in two stages: an energy limiting stage (Stage 1) and a falling rate stage (Stage 2) (Ritchie 1972) as indicated in Fig. 3. During stage 1, evaporation occurs at the maximum rate limited only by energy availability at the soil surface and therefore, K_r = 1. As the soil surface dries, the evaporation rate decreases below the potential evaporation rate (K_c max – K_{cb}). K_r becomes zero when no water is left for evaporation in the evaporation layer. Stage 1 holds until the cumulative depth of evaporation D_e is depleted which depends on the hydraulic properties of the upper soil. At the end of Stage 1 drying, D_e is equal to readily evaporable water (REW). REW ranges from 5 to 12 mm and highest for medium and fine textured soils (Table 1 of Allen et al., 2005). The evolution of K_r is presented in Fig. 3.

The second stage begins when D_e exceeds REW. Evaporation from the soil decreases in proportion to the amount of water remaining at the surface layer. Therefore reduction in evaporation during stage 2 is proportional to the cumulative evaporation from the surface soil layer as expressed in Eq. (16).

$$K_r = \frac{TEW - D_{e,j-1}}{TEW - REW} \text{ for } D_{e,j-1} > REW \tag{16}$$

where D_e, j-1 = cumulative depletion from the soil surface layer at the end of previous day (mm); The TEW and REW are in mm. The amount of water that can be removed by evaporation during a complete drying cycle is estimated as in Eq. (17).

$$TEW = 1000\left(\theta_{FC} - 0.5\theta_{WP}\right)Z_e \tag{17}$$

Where TEW =maximum depth of water that can be evaporated from the surface soil layer when the layer has been initially completely wetted (mm). θ_{FC} and θwp are in (m³ m⁻³) and Ze (m) = depth of the surface soil subject to evaporation. FAO-56 recommended values for Ze of 0.10-0.15m, with 0.10 m for coarse soils and 0.15 m for fine textured soils.

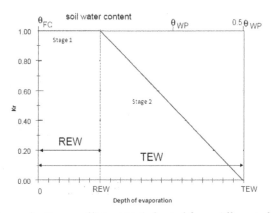

Fig. 3. Soil evaporation reduction coefficient K_r (adapted from Allen et al., 2005). REW stands for readily extractable water and TEW stands for total extractable water.

Calculation of K_e requires a daily water balance for the wetted and exposed fraction of the surface soil layer (f_{ew}). Eq. (18) is used to determine cumulative evaporation from the top soil layer (Allen et al., 2005).

$$D_{e,j} = D_{e,j-1} - \left(P_j - R_j\right) - \frac{I_j}{f_w} + \frac{E_j}{f_{ew}} + T_{ei,j} + D_{ei,j} \tag{18}$$

where $D_{e,j-1}$ and $D_{e,j}$ = cumulative depletion at the ends of days j-1 and j (mm); P_j and R_j = precipitation and runoff from the soil surface on day j (mm); I_j = irrigation on day j (mm); E_j = evaporation on day j (i.e., $E_j = K_e \times ET_o$) (mm); $T_{ei,j}$ = depth of transpiration from exposed and wetted fraction of the soil surface layer (f_{ew}) on day j (mm); and $D_{ei,j}$ = deep percolation from the soil surface layer on day j (mm) if soil water content exceeds field capacity (mm). Assuming that the surface layer is at field capacity following heavy rain or irrigation, the minimum value of $D_{e,j}$ is zero and limits imposed are $0 \leq D_{e,j} \leq TEW$. T_{ei} can be ignored except for shallow rooted crops (0.5-0.6m).

Evaporation is greater between plants exposed to sunlight and with air ventilation. The fraction of the soil surface from which most evaporation occurs is $f_{ew} = 1-f_c$.

$$f_{ew} = \min(1-f_c, f_w) \tag{19}$$

Where $1-f_c = 1-CC$; f_w is fraction of soil surface wetted by irrigation or rainfall; f_w is 1 for rainfall (Table 20 of Allen et al., 1998); f_c is fraction of soil surface covered by vegetation. In this study f_c is the canopy cover measured using *GreenSeekerTM*. Values of parameters used in the dual coefficient approach are presented in Table 1.

Parameter	Value
Field capacity, θ_{FC} (m^3 m^{-3})	30.1
Permanent wilting point, θ_{WP} (m^3 m^{-3})	15.0
Effective rooting depth, Z_r (m)	1.00
Depth of the surface soil layer, Z_e (m)	0.15
Total evaporable water, TEW (mm)	33.7
Readily evaporable water, REW (mm)	9
Total available water, TAW (mm)	160
Readily available water, RAW (mm)	96
The ratio of RAW to TAW, p (fraction)	0.6
Wetting fraction, f_w (fraction)	1

Table 1. The parameters of the soil used in the determination of K_s, K_e, and K_r in the FAO dual coefficient method.

The top soil layer (0-0.15 m) of the soil in this study is sandy clay loam. Readily extractable water (REW) is 9 mm for this soil texture (Table 1 of Allen et al., 2005). Field capacity and wilting point of this soil were determined as part of soil hydraulic properties characterization. Canola effective rooting depth was determined as part of National Brasicca Germaplasm Improvement Program (David Luckett, personal communication). Soil moisture content was monitored using on-site calibrated neutron probe. Soil moisture depletion fraction (p) of 0.6 m was taken from FAO-56 publication (Allen et al., 1998). Since the only source of water was rainfall, wetting fraction f_w of 1 was used.

4.4 AquaCrop approach of determining dual evapotranspiration coefficients

Eq. (11) gives evapotranspiration when the soil water is not limiting. When the soil evaporation and transpiration drops below their respective maximum rates, AquaCrop simulates ET$_a$ by multiplying the crop transpiration coefficient with the water stress coefficient for stomatal closure (Ks_{sto}), and the soil water evaporation coefficient with a reduction K_r [0-1] (Steduto et al., 2009) as

$$ET_a = (Ks_{sto}K_{cb} + K_rK_e) ET_o \tag{20}$$

AquaCrop calculates basal crop coefficient at any stage as a product of basal crop coefficient at mid-season stage $K_{cb(x)}$ and green canopy cover (CC). For canola $K_{cb(x)} = 0.95$ was used.

$$K_{cb} = K_{cb(x)} \times CC \tag{21}$$

$$K_e = K_{e(x)} \times (1-CC) \tag{22}$$

Evaporation from a fully wet soil surface is inversely proportional to the effective canopy cover. The proportional factor is the soil evaporation coefficient for fully wet and unshaded soil surface ($K_{e(x)}$) which is a program parameter with a default value of $K_{e(x)} = 1.1$ (Raes et al., 2009).

During the energy limiting (non-water limiting) stage of evaporation, maximum evaporation (E_x) is given by

$$E_x = K_e \, ET_o \; = [(1\text{-}CC)K_{ex}]ET_o \tag{23}$$

Where CC is green canopy cover; K_{ex} is soil evaporation coefficient for fully wet and non shaded soil surface (Steduto et al., 2009). In AquaCrop, K_{ex} is a program parameter with a default value of 1.10 (Allen et al., 1998). When the soil water is limiting, actual evaporation rate is given by

$$E_a = K_r E_x \tag{24}$$

Maximum crop transpiration (T_{rx}) for a well-watered crop is calculated as

$$T_{rx} = K_{cb} \, ET_o = [CC \, K_{cbx}]ET_o \tag{25}$$

K_{cbx} is the basal crop coefficient for well-watered soil and complete canopy cover.

5. Results and discussion

5.1 Soil water balance

The actual evapotranspiration determined using soil water balance method is presented in Table 2. Evapotranspiration was determined using Eq. (2) from measurement of 12 neutron probes several times during the season. Deep percolation and runoff were not measured. Therefore, values estimated by AquaCrop (Steduto et al., 2009; Raes et al., 2009) during the canola water productivity simulation were adopted.

DAP*	Rainfall (mm)	Deep percolation (mm)	Runoff (mm)	Change in storage (mm)	Evapotranspiration ET_a using water balance (mm)
0-13	6.5	0	0	-2.1	8.6
14-21	0	0	0	-1.8	1.8
22-28	36.9	4.6	0.5	13.4	18.4
29-35	23.4	24.6	1.4	-10	7.4
36-42	1.8	1.8	0	-3.1	3.1
43-49	6	2.2	0	-1.1	4.9
50-63	21.8	6.7	0	4.6	10.5
64-77	60	20.2	4.1	17.7	18
78-94	3.2	18.9	0	-25.6	9.9
95-118	58.7	21.2	1.6	6.7	29.2
119-143	81	34.3	3.8	-20.8	63.7
144-159	0	1.5	0	-39.6	38.1
160-173	103.9	8.6	14	30.3	51
174-196	31.6	3.8	0	-20.7	48.5
*DAP stands for days after planting				Seasonal	313

Table 2. Evapotranspiration determined using soil water balance method for canola planted on 30 April 2010 at Wagga Wagga (Australia).

The runoff estimated using AquaCrop was low, supporting the consensus that runoff from agricultural land is low. However, deep percolation past the 1.2 m was significant. The actual annual crop evapotranspiration estimated using this method was 313 mm. It can be observed that evapotranspiration was higher during the mid season and highly evaporative months.

5.2 Evapotranspiration coefficient

Single and dual evapotranspiration coefficients and crop canopy cover data are presented in Fig. 4. The K_c and K_{cb} values adopted from FAO-56 publication and adjusted for the local condition are shown in the Figure. The K_c and K_{cb} curves follow similar trend as the measured canopy cover curve. The canopy cover values were higher than the K_c and K_{cb} curves towards the end of the season. This is due to the fact that as an indeterminate crop, canola still had green canopy due to the ample rainfall during this late season stage of the crop. The soil evaporation coefficient K_e was correctly simulated using the top-layer soil water balance model. It can be seen that K_e is high during the initial and late season stages. It remained low and steady during the midseason stage. The higher number of K_e spikes are

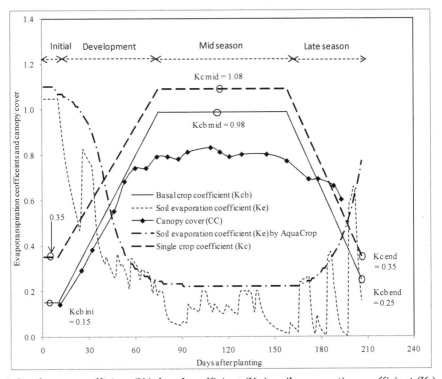

Fig. 4. Single crop coefficient (K_c), basal coefficient (K_{cb}), soil evaporation coefficient (K_e), crop canopy cover (CC) curves for canola having growth stage lengths of 10, 64, 84, and 48 days during initial, development, midseason, and late season stages. Indicated on curve are also single and basal crop coefficient (K_c and K_{cb}) at initial, midseason, and end of season stages. Day of planting is 30 April 2010.

due to frequent rainfall during the season. The K_e value estimated using AquaCrop followed similar trend to the manually calculated using Eq. (14). However, AquaCrop did not simulate response to individual rainfall events.

In the development stage, the soil surface covered by the crop gradually increases and the K_e value decreases. In the midseason stage, the soil surface covered by the crop reaches maximum and water loss is mainly by crop transpiration and K_e is as low as 0.05. In the late season stage, the K_e values are greater than that in the mid-season stage because of the senescence.

Evaporation and transpiration estimated using the dual coefficient approach (Fig. 5) are correctly simulated, with high evaporation during the initial and late stages, and low during the developmental and mid season stages. The fluctuation in the evaporation component is high at these stages and low and steady during the mid season stage except minor spikes after rainfall events. Evaporation during the late stage (late spring months) was high compared with the initial stage which is a winter period. The transpiration component was steady increasing during the crop development stage before reaching a maximum in late mid season stage and declined during the late season stage due to senescence. The trends in evaporation and transpiration were in perfect phase with the weather and crop phenology.

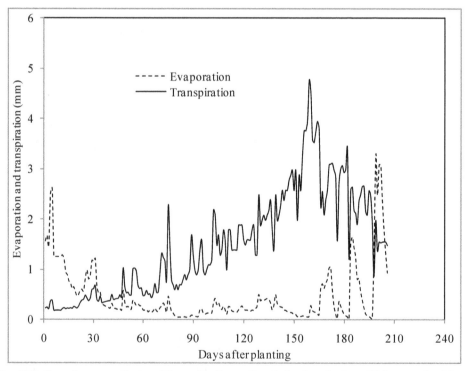

Fig. 5. Daily soil evaporation and transpiration estimated using dual coefficient method for canola planted on 30 April 2010 at Wagga Wagga, NSW (Australia).

Evapotranspiration varies during the growing period of a crop due to variation in crop canopy and climatic conditions (Allen et al., 1998). Variation in crop canopy changes the proportion of evaporation and transpiration components of evapotranspiration. The spikes in basal crop coefficient were high during the initial and crop development phases and decreases as the soil dries (Fig. 4). The spikes decrease as the canopy closes and much of ET is by transpiration. During the late season stage, there were fewer spikes because soil evaporation was low and almost constant. The largest difference between K_c and K_{cb} is found in the initial growth stage where evapotranspiration is predominantly in the form of soil evaporation and crop transpiration. Because crop canopies are near or at full ground cover during the mid-season stage, soil evaporation beneath the canopy has less effect on crop transpiration and the value of K_{cb} in the mid season stage is very close to K_c. Depending on the ground cover, the basal crop coefficient during the mid season stage may be only 0.05-0.10 lower than the K_c value. In this study $K_{cb\ mid}$ is 0.10 lower than $K_{c\ mid}$.

Some studies, carried out in different regions of the world, have compared the results obtained using the approach described by Allen et al. (1998) with those resulting from other methodologies. From this comparison, some limitations should be expected in the application of the dual crop coefficient FAO-56 approach. Dragoni et al. (2004), which measured actual transpiration in an apple orchard in cool, humid climate (New York, USA), showed a significant overestimation (over 15%) of basal crop coefficients by the FAO 56 method compared to measurements (sap flow). This suggests that dual crop coefficient method is more appropriate if there is substantial evaporation during the season and for incomplete cover and drip irrigation.

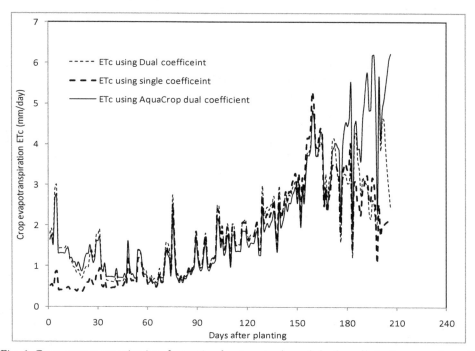

Fig. 6. Crop evapotranspiration determined using single and dual coefficient approaches of FAO 56 for a canola planted on 30 April 2010 at Wagga Wagga, NSW (Australia). ET_c estimated using AquaCrop (dual coefficient) is also presented.

Crop evapotranspiration estimated using single and double coefficients is presented in Fig. 6. ET_c estimated using AquaCrop is also presented in the Figure. It can be observed that ET_c estimated using the three approaches is similar except in the initial and late season stages. During the initial stage, the ET_c estimated using Eq. (14) and AquaCrop (Eqs. 21 and 22) are very close. However, the single coefficient method underestimated ET_c at this stage. During the initial stage when most of the soil is bare, evaporation is high especially if the soil is wet due to irrigation or rainfall. The single crop coefficient approach does not sufficiently take this into account. A similar pattern was observed during the late season stage. However, AquaCrop overestimated ET_c during this stage compared to the other two methods. The annual evapotranspiration estimated using different approaches was as follows: soil water balance (ET_a = 313 mm), single crop coefficient (ET_c = 332 mm), dual coefficient approach (ET_c = 366 mm with E of 79 mm and T of 288 mm), AquaCrop (ET_c = 382 mm with E of 139 mm and T of 243 mm). The evapotranspiration determined using soil water balance method is the "actual" evapotranspiration while the other methods measure potential evapotranspiration ET_c. Soil water depletion (Dr) in Eq. (6) was determined using soil moisture content measured during the season and it was found that Dr<RAW throughout the season indicating that there was no soil moisture stress (K_s = 1). That might be why the ET_c estimated using single coefficient method is close to the ET_c determined using soil water balance method. Approaches using dual coefficient (Eq. 14) and Eqs. (21 and 22) resulted in higher ET_c values. This might be due to the fact that in these approaches, the evaporation during the initial and late season stages was well simulated.

6. Conclusion

Two approaches of estimating crop evapotranspiration were demonstrated using a field crop grown in a semiarid environment of Australia. These approaches were the rootzone soil water balance and the crop coefficient methods. The components of rootzone water balance, except evapotranspiration, were measured/estimated. Evapotranspiration was calculated as an independent parameter in the soil water balance equation. Single crop coefficient and dual coefficient approaches were based on adjustment of the FAO 56 coefficients for local condition. AquaCrop was also used to estimate crop evapotranspiration using the dual coefficient approach. It was found that the dual coefficients, basal or transpiration coefficient K_{cb} and evaporation coefficient K_e, correctly depict the actual process. The effects of weather (rainfall and radiation) and crop phenology were correctly simulated in this method. However, single coefficient does not show the high evaporation component during the initial and late season stages. Generally, there is a strong agreement among different estimation methods except that the dual coefficient approach had better estimate during the initial and late season stages. The evapotranspiration estimated using different approaches was as follows: soil water balance (ET_a = 313 mm), single crop coefficient (ET_c = 332 mm), dual coefficient approach (ET_c = 366 mm with E of 79 mm and T of 288 mm), AquaCrop (ET_c = 382 mm with E of 139 mm and T of 243 mm). Evapotranspiration estimated using soil water balance method is actual evapotranspiration ET_a, while other methods estimate potential (maximum) evapotranspiration. Accordingly, ET estimated using rootzone water balance is lower than the ET estimated using the other methods. The single coefficient approach resulted in the lowest ET_c as it is not taking into account the evaporation spikes after rainfall during the initial and late season stages.

7. Acknowledgments

The senior author was research fellow at EH Graham Centre for Agricultural Innovation during this study. We also would like to thank David Luckett, Raymond Cowley, Peter Heffernan, David Roberts, and Peter Deane for professional and technical assistance.

8. References

Allen R.G., Pereira L.S., Raes D., Smith M. 1998. Crop evapotranspiration: guidelines for computing crop water requirements, FAO Irrigation and Drainage Paper 56., 300 p.

Allen R.G., Pereira L.S., Smith M., Raes D., Wright J.L. 2005. FAO-56 dual crop coefficient method for estimating evaporation from soil and application extensions. J Irrig Drain Eng ASCE, 131(1):2–13

Blaney, H.F. and Criddle, W.D. 1950. Determining water requirements in irrigated areas from climatological and irrigation data. USDA Soil Conserv. Serv. SCS-TP96. 44 pp.

Bonder, G., Loiskandl, W., Kaul, H.P. 2007. Cover crop evapotranspiration under semiarid conditions using FAO dual coefficient method with water stress compensation. Agric. Water Manag., 93 : 85-98.

Dragoni , D., Lakso, A.N., Piccioano, R.M. 2004. Transpiration of an apple orchard in a cool humid climate: measurement and modeling, Acta Horticulturae, 664:175-180.

Hawkins, R. H., Hjelmfelt, A. T., and Zevenbergen, A. W. 1985. Runoff probability, storm depth, and curve numbers. J. Irrig. Drain. Eng., 111(4): 330–340.

Hillel, D. 1997. Small scale irrigation for arid zones: Principles and options, Development monograph No. 2 , FAO, Rome.

Hillel, D. 1998. Environmental soil physics. Academic press. 771 pp. Elsevier (USA).

Monteith, J.L. 1981. Evaporation and surface temperature. Quart. J. Roy. Meteorol. Soc., 107:1-27.

Penman, H. L. 1948. "Natural evaporation from open water, bare soil and grass." Proc. Roy. Soc. London, A193, 120-146.

Penman, H.L. 1956. Estimating evaporation. Trans. Amer. Geoph. Union, 37:43-50.

Raes, D. 2009. *ET₀ Calculator*: a software program to calculate evapotranspiration from a reference surface. FAO Land Water Division. Digital Media Service No 36.

Raes, D., Steduto, P., Hsiao, T.C., Fereres, E., 2009. AquaCrop—The FAO crop model to simulate yield response to water: II. Main algorithms and soft ware description. Agron. J. 101:438–447.

Ritchie, J.T., 1972. Model for predicting evaporation from a row crop with incomplete cover. Water Resour. Res. 8, 1204–1213.

Riverina Development Australia, RDA (2011). Riverina – Food basket of Australia. Industry and Investment , NSW Government. accessed 30 July 2011.

Smith, M. 1992. CROPWAT, a computer program for irrigation planning and management. FAO Irrigation and Drainage Paper 46, FAO, Rome.

Steduto, P., Hsiao, T.C., Raes, D., Fereres, E., 2009. AquaCrop—the FAO crop model to simulate yield response to water. I. Concepts. Agron. J. 101:426-437.

Stern, H., de Hoedt, G., Ernst, J., 2000. Objective classification of Australian climates. Bureau of meteorology, Melbourne.

Suleiman A.A., Tojo Soler, C.M., Hoogenboom, G. 2007. Evaluation of FAO-56 crop coefficient procedures for deficit irrigation management of cotton in a humid climate. Agric. Water Maneg., 91:33-42.

Thornthwaite, C.W. 1948. An approach toward a rational classification of climate. Geograph. Rev., 38:55-94.

3

Assessment of Evapotranspiration in North Fluminense Region, Brazil, Using Modis Products and Sebal Algorithm

José Carlos Mendonça[1], Elias Fernandes de Sousa[2],
Romísio Geraldo Bouhid André[3], Bernardo Barbosa da Silva[4]
and Nelson de Jesus Ferreira[5]
[1]*Laboratório de Meteorologia (LAMET/UENF). Rod. Amaral Peixoto,
Av. Brennand s/n Imboassica, Macaé, RJ*
[2]*Laboratorio de Engenharia Agrícola (LEAG/UENF); Avenida Alberto Lamego,
CCTA, sl 209, Parque Califórnia, Campos dos Goytacazes, RJ*
[3]*Instituto Nacional de Meteorologia (INMET/MAPA); Eixo Monumental,
Via S1 – Sudoeste, Brasília, DF*
[4]*Departamento de Ciências Atmosféricas (DCA/UFCG); Avenida
Aprígio Veloso, Bodocongó, Campina Grande, PB*
[5]*Centro de Previsão de Tempo e Estudos Climáticos (CPTEC/INPE);
Av. dos Astronautas, Jardim da Granja, São José dos Campos, SP
Brazil*

1. Introduction

North Fluminense Region, Rio de Janeiro State, Brazil (Fig. 1) is known as a sugar cane producer. The production during harvest season 2007/08 were 4 million tons of sugar cane, that were transformed into 4.8 million sacks of sugar, 36,786 liters anhydrous alcohol (ethanol) and 91,008 liters of hydrated alcohol. Economically generated 250 million U. S. dollars (Morgado, 2009). However, this activity is declining in the region due to different factors, including hidric deficit and the use of irrigation techniques may reverse this situation(Azevedo et al., 2002). Some authors (Ide e Oliveira, 1986; Magalhães, 1987) define temperature as a factor of greater importance for sugar cane physiology maturation (ripening) because more the affecting nutrients and water absorption through transpiration flux is a non-controllable condition. Soil humidity is another preponderant factor to sugar cane physiology and varies in function of the cultivation cycle, development stage, climactic conditions and others factors, such as spare water in the soil. The soil moisture content varies during the growth that corresponds to the main cause of production variation. However, the precipitation distribution along the year and spare soil water for the plant disposition are more important in the vegetative cycle of the sugar cane that total precipitation. (Magalhães, 1987).

The physical properties of energy exchange between the plant community and environment such as momentum, latent heat, sensible heat and others are evidenced by the influence they exert on physiological processes of plants and the occurrence of pests and diseases, which affect the productive potential of plants species exploited economically (Frota, 1978). The radiation components measurements of energy balance in field conditions have direct applicability in agricultural practices, especially in irrigation rational planning, appropriate use of land in regional agricultural zoning, weather variations impact on agricultural crops, protecting plants, among others. The knowledge advance in micro-scale weather, as well as the instrumental monitoring technology evolution has allowed a research increase in this area. Energy balance studies on a natural surface based on energy conservation principle. By accounting means for components that make up this balance, can be evaluate the net radiation plots used for the flow of sensible and latent heat.

The analysis of data collected by artificial satellites orbiting planet earth, allows the determination of various physical properties of planet, consequently, spatial and temporal modifications of different ecosystems are able to be identified.

According Moran et al. (1989), estimative of evepotranspiration – ET, based in data collected in meteorological stations have the limitation of representing punctual values that are capable of satisfactory representing local conditions but, if the objective is to obtain analysis of a regional variation of ET using a method with interpolation and extrapolation from micro-meteorological parameters of an specific area, these punctual data may increase the uncertainty of the analysis.

Trying to reduce such uncertainty degree, different algorithms were developed during the last decades to estimate surface energy flux based in the use of remote sensing techniques.

Bastiaanssen (1995) developed the 'Surface Energy Balance Algorithm for Land - SEBAL', with its validation performed in experimental campaigns in Spain and Egypt (arid climate) using Landsat 5 –TM images. This model involves the spatial variability of the most agro-meteorological variables and can be applied to various ecosystems and requires spatial distributed visible, near-infrared and thermal infrared data together with routine weather data. The algorithm computes net radiation flux – Rn, sensible heat flux - H and soil heat flux - G for every pixel of a satellite image and latent heat flux - LE is acquired as a residual in energy balance equation (Equation 01). This is accomplished by firt computing the surface radiation balance, flowed by the surface energy balance. Althoygh SEBAL has been designed to calculate the energy partition at the regional scale with minimum ground data (Teixeira, 2008).

Roerink et al. (1997) also used Landsat 5 –TM images to evaluate irrigation's performance in Argentina and AVHRR/NOAA sensor images in Pakistan. Combination of Landsat 5 – TM and NOAA/AVHRR images were used by Timmermans and Meijerink (1999) in Africa. Latter, Hafeez et al. (2002) used the SEBAL algorithm with the ASTER sensor installed onboard 'Terra' satellite while studying Pumpanga river region in Philippines. These authors concluded that the combination of the high spatial resolution of ETM+ and ASTER sensors, together with the high temporal resolution from AVHRR and MODIS, provided high precision results of water balance and water use studies on regional scale.

In Brazil, several research center are conducting research using the SEBAL algorithm specially 'Federal University of Campina Grande, PB - UFCG', 'National Institute of Space Research - INPE' and others.

Sebal was developed and validated in arid locations and one of its peculiarities is the use of two anchors pixels (hot pixel – LE = 0 and cold pixel – H =0) with the determination or selection of hot pixel easier in dry climates. In humid and sub-humid climates is not easy determine a hot pixel, where the latent heat flux is zero or null.

The objectives of the research described in this work are (i) to evaluate two propositions to estimate the sensible heat flux (H) and (ii) to evaluate two methods for conversion of ETinst values to ET24h on the daily evepotranspiration to estimate evepotranspiration in regional scale using SEBAL algorithm, MODIS images, the two propositions to estimate H and meteorological data of the four surface meteorological stations.

2. Materials and methods

2.1 Study area

The Norte Fluminense region in Rio de Janeiro State, Brazil, has an area of 9.755,1 km², corresponding to 22% of the state's total area. Among its agricultural production, sugar cane plantations are predominant as well as cattle production. In the last years irrigation technologies for fruit production are being promoted and implemented by the government. Nowadays, passion fruit, guava, coconut and pineapple plantations extend for more than 4.000 ha (SEAAPI, 2006).

According Koppen, this region's clime is classified as Aw, that is, tropical humid with rainy summers, dry winters and temperatures average above 18 °C during the coolest months. The annual mean temperatures are of 24°C, with a little thermal amplitude and mean rain precipitation values of 1.023 mm (Gomes, 1999).

The area under study is showed in Figure 1, comparing the area of the Norte Fluminense region within the Rio de Janeiro state and the RJ state within Brazil.

Fig. 1. Study area localization.

2.2 Digital orbital images – MODIS images

Daily MOD09 and MYD09 data (Surface Reflectance – GHK / 500 m and GQK / 250 m) and MOD11A1 and MYD11A1 data (Surface Temperature - LST) were used in this research, totalizing 24 scenes over the 'tile' h14/v11 corresponding to Julian Day 218th, 227th, 230th, 241st, 255th, 285th, 320th and 339th in 2005 and 15th, 36th, 63rd , 102nd, 116th, 139th, 166th, 186th, 189th, 190th, 191st, 200th, 201st, 205th, 208th and 221st in 2006. These days were selected because no cloud covering was registered over the study area during the satellite's course over the area were obtained from the Land Processes Distributed Active Archive Center (LP-DAAC), of the National Aeronautics and Space Administration (NASA), at http://edcimswww.cr.usgs.gov/pub/imswelcome/.

The GHK – 500 m (Blue, Green, Red, Nir, Mir, Fir, Xir) reflectance band were resampled fron 500 m to 250 m. The Red and Nir bands were excluded and GQK (250 m) bands included. This operation aimed to input the value of the red and nir bands in the algorithm. The LST bands were also resampled from 1000 m to 250 m.

The software Erdas Image – Pro, version 8.7 was used for the piles, compositions, clippings and algebra. The Model Maker tool was used to application of the algorithm and the thematic maps were produced using the software ArcGis 9.0.

2.3 Meteorological data

Surface data were collected in two micro-meteorological stations from the Universidade Estadual do Norte Fluminense – UENF, installed over agricultural areas cultivated with sugar cane (geographical coordinates: 21° 43′ 21,8″ S and 41° 24′ 26,1″ W), and 'dwarf green' coconut irrigated (geographical coordinates: 21° 48′ 31,2″ S and 41° 10′ 46,2″ W).

The micrometeorological stations installed in both areas (sugar cane and coconut) were equipped with the following sensor: 1 Net radiometer NR Lite (Kipp and Zonen), 2 Piranometer LI 200 (Li-Cor), 2 Probe HMP45C-L (Vaissala), 2 Met One Anemometer (RN Yong) and 3 HFP01SC_L Soil Healt Flux Plat (Hukseflux). All data from were collected every minute and average values extracted and stores every 15 min in a datalogger CR21X (Sugar cane) and CR 1000 (coconut). Both dataloggers are Campbell Scientific's (USA). The horizontal bars were placed 0.50 m above crop canopy (first level) and 2.0 m between the first and second bars. This standard was maintained all crop cycle and bars relocated where necessary (sugar cane station). In coconut station the relocated was not necessary.

These stations were installed in the center of an area of 5,000 hectare (sugar cane – Santa Cruz Agroindustry) 256 hectare (coconut – Agriculture Taí).

Fig. 2. Localization of the surface micro-meteorological and meteorological stations installed in the study area.

The meteorological stations, both installed on grass (*Paspalum Notatum L.*) are property of research center. The Thies Clima model (Germany) installed at the UENF's Evapotranspiration Station – Pesagro Research Center, (geographical coordinates: 21° 24' 48" S and 41° 44' 48" W) is an automatic station. Is equipped with 1 Anemometer, 1 Barometer, 1 Termohygrometer, 1 Piranometer and 1 Pluviometer. All sensor are connected to a datalogger model DL 12 – V. 2.00 – Thies Clima, recording values every minute and stored an average every 10 minutes.

The Agrosystem model install at the Meteorological Station of the Experimental Campus 'Dr. Leonel Miranda' – UFRRJ, (geographical coordinates: 21° 17' 36" S and 41° 48' 09" W) contains 1 Anemometer, 1 Barometer, 1 Termohygrometer, 1 Piranometer and 1 Pluviometer and recording values every minute and stored an average every 10 minutes.

All geographical coordinates are related to Datum WGS 84 – zone 24, with average altitude of 11 m. The localization of the surface stations, where meteorological data used in this study were collected are showed in Figure 2.

2.4 Real evapotranspiration estimation with SEBAL

To calculate surface radiation balance was used the Model Maker tool from the software Erdas Image 8.6. The estimations of the incident solar radiation and the long wave radiation emitted by the atmosphere to the surface were performed in electronic sheet.

To better understand the different phases of the Sebal algorithm using Modis products, a general diagram of the computational routines are shown in Figure 3.

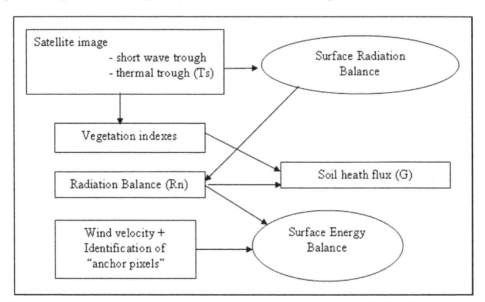

Fig. 3. Diagram of the computational routines for determination of the Surface Energy Balance using SEBAL, form MODIS products. (Modified from Trezza (2002).

A schematic diagram for the estimation of the surface radiation balance (Rn), adapted to MODIS images is showed in Figure 4.

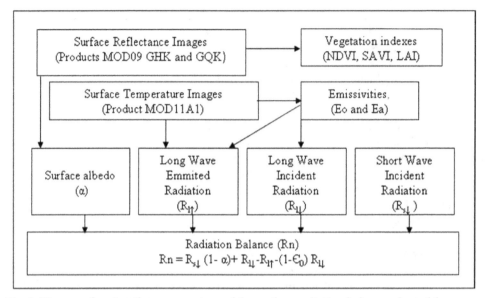

Fig. 4. Diagram showing the process steps of the surface radiation balance adapted for MODIS images.

Detailed processes, as well as the equations for the SEBAL algorithm development, may be obtained in Bastiaanssen et al. (1998). In the present work two propositions were assumed to select the anchor pixels, the first was similar to the one used by Bastiaanssen (1995), with the selection of two pixels with external temperatures (hot pixel/LE = 0 and cool pixel/H = 0). The hot pixel always comprising an area of exposed soil with little vegetation and the cool pixel localized in the interior of a great extension water body. The first proposition was called as 'H_Classic'.

With the hypothesis that the linear relation dT = a + d.Ts would be better represented with the selection of a hot pixel with its energy balance components previously known, specially the sensible heat flux (H) and in regions of humid and sub-humid climate be difficult identifying de hot pixels, which can hardly meet the condition of being dry, or have LE = 0, the second hypothesis was formulated. The criterion used for the selection of the cool pixel was the same as in the first hypothesis, that is, to be localized inside a water body of a great extension, but the selection of the hot pixel, where determination of the H values estimate as residue of the Penman-Monteih FAO56 equation using meteorological data from installed at the UENF's Evapotranspiration Station – Pesagro Research Center. This second hypothesis was called 'H_Pesagro'.

2.5 Latent heat flux (LE)

Latent heat flux (vapor transference to the atmosphere trough the process of vegetal transpiration and soil water evaporation) was computed by the simple difference between the radiation balance cards, soil heat flux and sensible heat flux:

$$LE = Rn - G - H \qquad (1)$$

where: LE represents the latent heat flux, Rn is the radiation balance and G is the soil heat flux, all expressed in W m^{-2} and obtained during the course of the satellite over the study area.

The value of the instantaneously latent heat flux (LE_{inst}), integrated at the time (hour) of the satellites passage (mm h^{-1}) is:

$$LE_{inst} = 3600 \frac{LE}{\lambda} \qquad (2)$$

where: LE_{inst} is the value of instantaneously ET, expressed in mm h^{-1}; LE is the latent heat flux at the moment of the sensor's course and λ is the water vaporization latent heat, expressed by the equation:

$$\lambda = 2,501 - 0,00236 \, (Ts - 273,16) * 10^6 \qquad (3)$$

where: Ts is the surface temperature chart (ºC) obtained by the product MOD11A1 (K).

With the radiation balance, soil heat flux and latent heat flux charts, the evaporative fraction was obtained and expressed by the equation:

$$\Lambda = \frac{LET}{Rn - G} \qquad (4)$$

The evaporative fraction has an important characteristic, it regularity and constancy in clear sky days. In this sense, we can admit that its instantaneously character represents its diurnal mean value satisfactorily, enabling the estimation of daily evapotranspiration by the equation:

$$ET_{24h} = \frac{86400 \, \Lambda \, Rn_{24h}}{\lambda} \qquad (5)$$

where: Rn_{24h}, is the mean radiation balance occurred during a period of 24 h, expressed in W.m^{-2}, obtained by the equation:

$$Rn_{24h} = (1 - \alpha) \, Rs24h - 110 \, \tau_{sw} 24h \qquad (6)$$

where: α, is the surface albedo; Rs_{24h}, is the daily mean radiation of short incident wave expressed in W m^{-2} and $\tau_{sw} 24h$, is the mean daily atmospheric transmissivity.

To determine Rs_{24h} values, an approximation similar to the method proposed by Lagouarde and Brunet (1983) for the estimation of diurnal cycles of Rn and Rs↓ in clear sky days, was used. With the values of Rn_{24h}, Rs_{24h} and the surface albedo, extracted from the PESAGRO pixel, a linear regression between these values was performed to obtain a regression equation, its coefficients a_1 and b_1 and then to calculate the Rn_{24h} chart as a function of the short wave balance. To determine the linear regression the following equation was used:

$$Rn24h = a_1 (1 - \alpha) * Rs24h + b_1 \qquad (7)$$

Allen et al. (2002) defined the evaporative fraction of reference (ETrF) as the relation between the ET_{inst} chart and the ETo integrated at the same moment and computed with data obtained from a meteorological station, that is:

$$ETrF = \frac{ET_{inst}}{ET_{FAO56}} \tag{8}$$

This procedure generates a type of hourly-cultive coefficient (kc_h), admitting that this relation represents the daily relation expressed by the equation:

$$Kc\_h = \frac{ETinst}{EToh} = \frac{ET24}{ETo24} \tag{9}$$

Admitting the relation represented in equation 09 it is possible to obtain the ET_{24h} expressed in mm day[-1] from the equation:

$$ET_{24h} = ETrF * ETo_{24} \tag{10}$$

In the present work, four values of ET24h$_{SEBAL}$ were estimated for the same day, applying equations 5 and 10 to the 'H_Classic' and H_Pesagro' propositions.

3. Results and discusion

3.1 Daily evapotranspiration (ET$_{24h}$)
3.1.1 Determination of Rn24h values
To determine Rn24h charts, an adaptation proposed by Ataide (2006) for the sinusoidal model estimator of the cycle of radiation balance for clear sky days, based in an approximation similar to the Lagourade and Brunet (1983) method, was adopted.

Looking forward for reliability and applicability in the generation of the Rn24h charts form values of Rs↓24h, a linear regression between the short wave balance and the daily radiation balance was performed, where the regression equation coefficients were determined as $a =$ 0,9111 and $b = -23,918$.

The coefficients obtained (a and b) are next to the values found by Alados et al. (2003), whit values of $a = 0,709$ and $b = -25,4$ where values of global solar radiation (Rg) and not short wave balance (BOC) were used in the linear regression, thus excluding the effect of the surface albedo in the calculation. Considering that values of Rg were determined in a standard meteorological station, installed on a grass field, with values of albedo varying between 20 and 25 %, the coefficients determined by the linear regression between values of BOC and Rn24h tent to be in agreement with the values mentioned by Alados et al. (2003).

Thus, the radiation balance for the daily period (Rn24h) was ultimately determined for each pixel of the study scene by the equation:

$$Rn_{24h} = 0,9111* (1 - \text{chart of albedo}) * Rs{\downarrow}24h -23,918 \tag{11}$$

3.1.2 Determination of the ET24h values
Based on charts of Rn, G, H, LE, Ts and α and values of ETo$_{24h}$ and ETo$_{inst}$, estimated from data observed at Pesagro's meteorological station, four values of ET24h were estimated for each scene studied: ET24h_'Classic' w/ETrF; ET24h_'Classic' w/Rn24h; ET24h_'H_Pesagro' w/ETrf and ET24h_'H_Pesagro' w/Rn24h.

Mean, maximum and minimum values obtained in charts of daily evapotranspiration (ET24h) estimated with the 'H_Classic' proposition and expressed in mm day[-1], are showed in Table 1.

DJ	Mean		Maximum		Minimum	
	Rn 24h	ETr_F	Rn 24h	ETr_F	Rn 24h	ETr_F
218	3,42	4,25	6,51	12,69	0,0	0,0
227	2,88	2,89	6,89	7,64	0,0	0,0
230	3,13	3,19	6,99	7,76	0,0	0,0
241	3,25	2,98	7,39	7,48	0,0	-1,04
255	4,07	3,64	8,25	8,27	0,0	-0,10
285	4,82	4,17	9,63	9,82	0,0	-0,60
320	4,50	3,70	10,65	10,15	0,0	-1,10
339	5,25	4,52	10,75	10,37	0,0	-0,81
15	4,77	4,06	10,91	10,97	0,0	-2,17
36	4,65	4,16	10,12	10,34	0,0	-1,64
63	5,10	5,67	9,40	11,82	0,0	-0,37
102	4,06	3,67	7,75	8,23	0,0	-1,10
116	3,27	3,20	6,93	7,65	0,0	-2,22
139	2,78	2,71	5,94	6,62	0,0	-0,67
166	2,73	2,88	5,45	6,73	0,0	-0,43
186	2,16	2,47	5,48	6,93	0,0	-0,83
189	2,75	2,99	5,61	6,90	0,0	-0,13
190	3,09	2,71	7,23	7,08	0,0	-0,20
191	2,27	2,56	5,68	7,23	0,0	-1,23
200	2,02	2,28	5,51	7,11	0,0	-0,52
201	2,87	3,46	5,84	8,31	0,0	-0,02
205	3,36	4,03	5,84	8,31	1,05	1,07
208	2,54	3,09	6,09	8,26	0,0	-0,98
221	2,88	3,08	6,59	7,82	0,0	-1,25

Table 1. Statistical data of daily evapotranspiration charts (ET24h) of the study area using the 'H_Classic' proposition w/ Rn24h and w/ ETr_F, in mm day[-1].

Average mean data showed in Table 1 are similar, with a slight superiority for the values estimated by the method using Rn24h for the ET estimative. Minimum values for ETr_F have negative values. Tasumi et al. (2003), using SEBAL in Idaho, U.S.A., also observed negative values for ET and attributed such results to systematic errors caused by diverse parameterizations used during the process of energy balance estimation.

Average mean, maximum and minimum values obtained in charts of daily evapotranspiration (ET24h) estimated with the "H_Pesagro' proposition, expressed in mm day[-1], are showed in Table 2.

	Mean		Maximum		Minimum	
DJ	Rn 24h	ETr_F	Rn 24h	ETr_F	Rn 24h	ETr_F
218	4,45	5,34	6,51	13,38	2,39	2,18
227	4,65	4,61	6,89	7,71	1,87	1,53
230	4,83	4,86	6,99	7,75	0,78	0,60
241	5,84	5,26	7,44	7,53	4,44	3,45
255	6,00	5,30	8,26	8,28	3,78	2,79
285	7,29	6,21	9,75	9,94	5,17	3,82
320	7,31	5,84	10,69	10,39	4,64	3,14
339	7,01	5,99	10,81	10,70	2,17	1,40
15	7,92	6,51	10,96	11,02	2,60	1,48
36	8,12	7,01	10,23	10,84	5,79	4,24
63	6,69	7,35	9,46	12,07	3,90	3,60
102	5,41	4,85	7,75	8,23	0,80	0,50
116	4,62	4,56	6,95	7,65	0,0	-0,46
139	4,27	4,13	5,95	6,62	2,60	2,21
166	4,27	4,11	5,95	6,62	2,61	2,21
186	3,37	3,84	5,48	7,12	1,73	1,75
189	3,86	4,16	5,62	6,94	2,28	2,14
190	5,49	4,74	7,25	7,10	4,41	3,39
191	3,74	4,16	5,70	7,25	0,36	0,27
200	3,29	3,70	5,58	7,20	1,69	1,74
201	3,36	4,03	5,83	8,31	1,05	1,08
205	4,50	5,00	5,95	7,91	3,40	3,26
208	4,33	5,11	6,09	8,35	2,93	3,06
221	4,68	4,93	6,62	7,86	2,88	2,65

Table 2. Statistical data of daily evapotranspiration charts (ET 24h) of the study area using the 'H_Pesagro' proposition w/ Rn 24hs and w/ ETr_F, in mm day^{-1}.

Average mean values of the same magnitude order and with a slight superiority to values estimated using Rn24h are obse4rved in Table 2. In a general way, by the use of the 'Classic' proposal as well as by 'Pesagro' proposal, a higher amplitude of the estimated values is observed when using the method of ETr_F.

Values of ET 24h$_{SEBAL}$, observed in pixels where the micro-meteorological and meteorological stations were located (pixels from Pesagro, UFFRJ, Sugar-cane and Coconut), were correlated with values of ETo estimated by the equation of Penman-Monteith_FAO (ETo PM_FAO56) with data observed in Pesagro Station. Figures 5, 6, 7 and 8 show graphical representations of the regression analysis, the adjustment equation and the correlation coefficient (R^2), obtained among the values estimated by SEBAL for all four methods used.

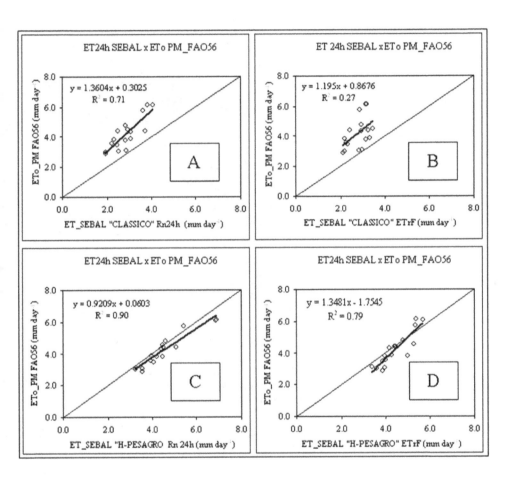

Fig. 5. Correlation between values of ET24h estimated with the method FAO (PM_FAO56) with data collected at PESAGRO station and values of ET24h estimated by SEBAL with propositions "H_Classic" w/Rn24h (A), "H_Classic" w/ETr_F (B), "H_Pesagro" w/Rn24h (C) and "H_Pesagro" w/ETr_F (D) observed in pixel from Pesagro, expressed in mm day^{-1}.

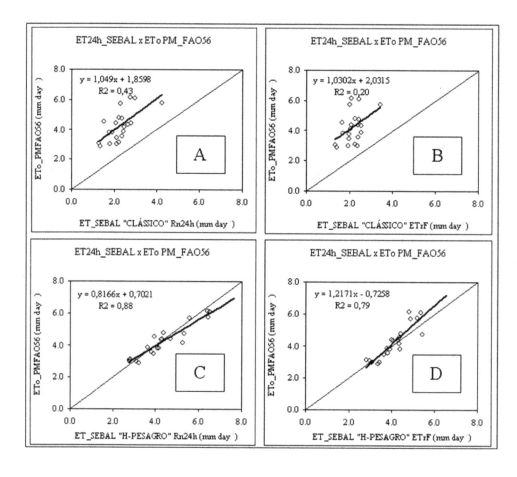

Fig. 6. Correlation between values of ET24h estimated with the method FAO (PM_FAO56) with data collected in PESAGRO station and values of ET24h estimated by SEBAL with propositions "H_Classic" w/Rn24h (A), "H_Classic" w/ETr_F (B), "H_Pesagro" w/Rn24h (C) and "H_Pesagro" w/ETr_F (D) observed in pixel pixel from UFRRJ, expressed in mm day[-1].

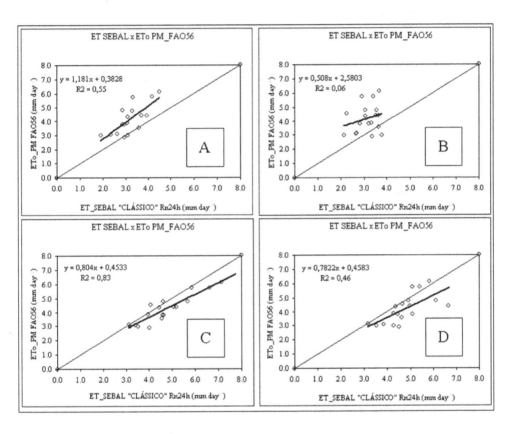

Fig. 7. Correlation between values of ET24h estimated by the method FAO (PM_FAO56) with data collected from PESAGRO station and values of ET24h estimated by SEBAL with propositions "H_Classic" w/Rn24h (A), "H_Classic" w/ETr_F (B), "H_Pesagro" w/Rn24h (C) and "H_Pesagro" w/ETr_F (D) observed in pixel from Sugar-cane (SANTA CRUZ AGROINDUSTRY), expressed in mm day[-1].

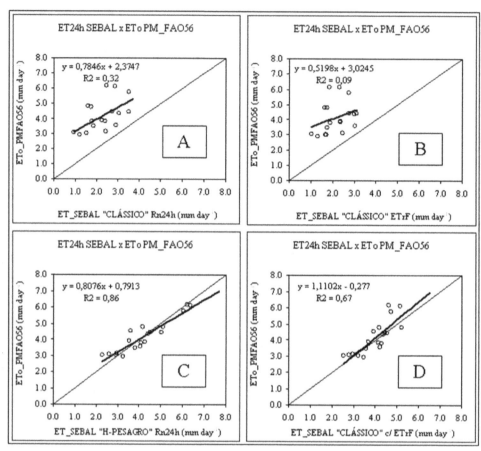

Fig. 8. Correlation between values of ET24h estimated by the method FAO (PM_FAO56) with data collected from PESAGRO station and values of ET24h estimated by SEBAL with propositions "H_Classic" w/Rn24h (A), "H_Classic" w/ETr_F (B), "H_Pesagro" w/Rn24h (C) and "H_Pesagro" w/ETr_F (D) observed in pixel from Coconut (**AGRICULTURE TAÍ**) expressed in mm day-1.

Observing Figures 5, 6, 7, and 8, it is possible to conclude that the proposition 'H_Classic' under estimated values projected by PM_FAO56 method, showing better results for values estimated using Rn24h.

Proposition 'H_Pesagro', although in a slight way, super estimated values of the ETo estimated with data from the meteorological station Pesagro, in all four control points, showing higher correction coefficients than the others with emphasis for the method using Rn24h.

Hafeez et al. (2002) applied SEBAL using MODIS images in Philippines and observed that the ET_SEBAL super estimated in 13,5 % the values of ETo estimated by PM_FAO56, justifying such behavior due to the spatial resolution of 1.000 m of the surface temperature chart (MOD11A1).

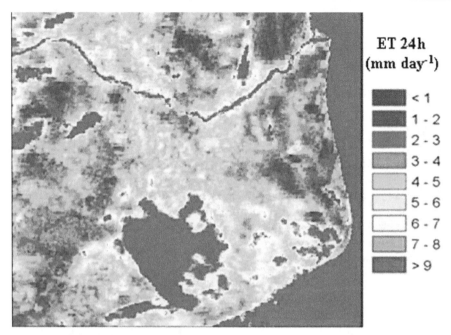

Fig. 9. Images of the daily evapotranspiration for the dry period in the Fluminense North Region, Rio de Janeiro State. DJ 2005218.

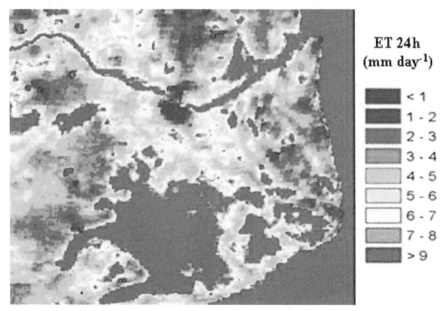

Fig. 10. Images of the daily evapotranspiration for the humid period in the Fluminense North Region, Rio de Janeiro State. DJ 2006015.

Allen et al. (2001), using images of LANDSAT in the basin of river Bear, North-East region of the U.S.A., observed that SEBAL showed a good precision for the estimation of ET, compared with weighing lysimeters, super estimating monthly mean values in 16% and 4 % for seasonal values.

Images of the daily evapotranspiration for the dry and humid periods in the Fluminense North Region, Rio de Janeiro State is showed in Figures 9 (DJ 2005218) and 10. (DJ 2006015).

4. Conclusion

In accordance with the proposed objectives in this work, it is possible to conclude that in conditions de sub-humid climate: For the estimative of sensible heath flux, the use of proposition 'H_Pesagro' resulted more efficient than 'H_Classic'; The method that uses values of mean radiation balance integrated in 24 hours (Rn24h) is more consistent than the method that uses the reference evaporative fraction (ETr_F) for the conversion of instantaneous evapotranspiration values (ET_{inst}) in daily values (ET_{24h}).

5. Acknowledgements

The authors are grateful for the National Counsel for Scientific and Technological Development – CNPq and the Coordenação de Aperfeiçoamento de Pessoal de Nível Superior – CAPES, for the financial support and logistics that made this study possible.

6. References

Alados, C.L.; Pueyo, Y.; Giner, M.L.; Navarro, T.; Escos, J.; Barroso, F.; Cabezudo, B.; Emlen, G.M., 2003. Quantitative characterization of the regressive ecological succession by fractal analysis of plant spatial patterns. Ecological Modell. v.163, p.1-17.

Allen, R. G.; Pereira, L. S.; Raes, D.; Smith, M., 1998. Crop evapotranspiration – Guidelines for computing crop water requeriments. FAO Irrigation and Drainage Paper 56, Rome, Italy, 318 p.

Allen, R.G.; Tasumi, M.; Trezza, R.; Bastiaanssen, W.G.M., 2002. SEBAL - Surface Energy Balance Algorithms for Land. Advanced training and users manual, Version 1.0. University of Idaho, EUA. 97 p.

Ataíde, K.R.P., 2006. Determinação do saldo de radiação e radiação solar global com produtos do sensor MODIS Terra e Aqua. Tese (Mestrado em Meteorologia) - Campina Grande, PB - Universidade Federal de Campina Grande – UFCG, 88p.

Azevedo, H.J.; Silva Neto, R.; Carvalho, A. M.; Viana, J.L.; Mansur, A.F.U., 2002. Uma análise da cadeia produtiva da cana-de-açúcar na Região Norte Fluminense. Observatório sócio-econômico da Região Norte Fluminense – Boletim Técnico n° 6, 51p.

Bastiaanssen, W.G.M., 1995. Regionalization of surface flux densities and moisture indicators in composite terrain. Ph,D Thesis, Wageningen Agricultural University, Wageningen, The Netherlands. 273p.

Bastiaanssen, W.G.M.; Pelgrum, H.; Wang, J.; Ma, Y.; Moreno, J.; Roerink, G. J.; van der Val, T., 1998. A remote sensing surface energy balance algorithm for land (SEBAL):Part 2 validation, Journal of Hidrology, v, 212-213: 213-229.

Frota, P.C.E., 1978. Estudo do calor sensível e latente no interior de uma cultura de milho (Zea mays L.), Dissertação (Mestrado em Agrometeorologia), Piracicaba, SP, Universidade Luiz de Queiroz - ESALQ/USP, 75p.

Gomes, M.C.R., 1999. Efeito da irrigação suplementar na produtividade da cana-de-açúcar em Campos dos Goytacazes, RJ. Dissertação (Mestrado em Produção Vegetal) - Campos dos Goytacazes - RJ, Universidade Estadual do Norte Fluminense - UENF, 51p.

Hafeez, M.M.; Chemin, Y,; Van de Giesen, N.; Bouman, B.A.M., 2002. Field evapotranspiration estimation in Central Luzón, Philippines, using different sensors: Landsat 7 ETM+, Terra Modis and Aster. Symposium on Geospatial Theory, Processing and Applications. Ottawa-Canadian. 7 p.

Ide, B. Y.; Oliveira, M.A. de, 1986. Efeito do clima na produção da cana-de-açúcar. In: Seminário de Tecnologia Agronômica, 3. Piracicaba, SP. Anais.... São Paulo: COPERSUCAR, p.573-583.

Lagouarde, J.P.; Brunet, Y., 1983. A simple model for estimating the daily upward long wave surface radiation flux from NOOA/AVHRR data. International Journal of Remote Sensing. 14(5):907-925.

Land Processes Distributed Active Archive Center - LP-DACC Available online: http://edcimswww.cr.usgs.gov/pub/imswelcome/ (Accessed on 15 April/2005).

Magalhães, A. C. N. 1987. Ecofisiologia da cana-de-açúcar: aspectos do metabolismo do carbono na planta. In: Castro, P.R.C.; Ferreira, S.O.; Yamada, T. (Ed.). Ecofisiologia da Produção Agrícola. Piracicaba, SP: Potafós, p. 113-118.

Moran, M. S.; Jackson, R. D.; Raymond L.; Gay, L.; Slater, P., 1999. Mapping surface energy balance components by combining Landsat Thematic Mapper and groudbase meteorological data. Remote Sensing of Environment. n.30:77-87.

Morgado, I.F., 2009. Agroindústria Sucroalcooleira do Estado do Rio de Janeiro. Universidade Cândido Mendes – UCAM. Available online: http://www.infoagro.ucam-campos.br/agro_in_rio.htm. (Accessed on 25 June/2010).

Paiva, C.M.; Liu, W.T.H.; Franca, G.B.; Filho, O.C. R., 2004. Estimativa das componentes do balanço de energia via satélite através do modelo SEBAL. XIII Congresso Brasileiro de Meteorologia, Fortaleza, CE. Anais.

Roerink, G.J.; Bastiaanssen, W.G.M.; Chambouleyron, J.; Menenti, M., 1997. Relating crop water consumption to irrigation water supply to remote sensing. Water Resources Management. 11: 445-465.

Secretaria Estadual de Agricultura, Pesca e Desenvolvimento do Interior - SEAAPI, RJ (2006). Available online: www.seaapi.rj.gov.br/frutificar. (Accessed on 15 November/ 2006).

Silva, B.B.; Lopes, G.M.; Azevedo, P.V., 2005. Balanço de radiação em áreas irrigadas utilizando imagens Landsat 5 –TM. Revista Brasileira de Meteorologia. V.20 (2): 243-252.

Tasumi, M., 2003. Progess in operational estimation of regional evapotranspiration using satellite imagery. PhD Dissertation. Idaho State University. Idaho. USA. 379 p.

Teixeira, A. H. C., 2008. Measurements and modelling of evepotranspiration to assess agricultural water productivity in basins with changing land use patterns - a case study in the São Francisco River basin, Brazil. PhD Dissertation. Wageningen University.. Nederland. 239 p.

Timmermans, W.J.; Meijerink, A.M.J., 1999. Remotely sensed actual evapotranspiration: implications for groundwater management in Botzwana. Journal of Applied Geohydrology. 1:222-233.

Trezza, R., 2002. Evapotranspiration using a satellite-based energy balance with sandarized ground control. PhD Dissertation. Utah State University. Logan. USA. 247p.

4

Hargreaves and Other Reduced-Set Methods for Calculating Evapotranspiration

Shakib Shahidian[1], Ricardo Serralheiro[1], João Serrano[1],
José Teixeira[2], Naim Haie[3] and Francisco Santos[1]
[1]University of Évora/ICAAM
[2]Instituto Superior de Agronomia
[3]Universidade do Minho
Portugal

1. Introduction

Globally, irrigation is the main user of fresh water, and with the growing scarcity of this essential natural resource, it is becoming increasingly important to maximize efficiency of water usage. This implies proper management of irrigation and control of application depths in order to apply water effectively according to crop needs. Daily calculation of the Reference Potential Evapotranspiration (ETo) is an important tool in determining the water needs of different crops. The United Nations Food and Agriculture Organization (FAO) has adopted the Penman-Monteith method as a global standard for estimating ETo from four meteorological data (temperature, wind speed, radiation and relative humidity), with details presented in the Irrigation and Drainage Paper no. 56 (Allen et al., 1998), referred to hereafter as PM:

$$ET_o = \frac{0.408\Delta(R_n - G) + \gamma \dfrac{900}{T + 273} u_2 (e_s - e_a)}{\Delta + \gamma(1 + 0.34u_2)} \tag{1}$$

where:
R_n – net radiation at crop surface [MJ m^{-2} day^{-1}],
G – soil heat flux density [MJ m^{-2} day^{-1}],
T – air temperature at 2 m height [°C],
u_2 – wind speed at 2 m height [m s^{-1}],
e_s – saturation vapor pressure [kPa],
e_a – actual vapor pressure [kPa],
e_s-e_a – saturation vapor pressure deficit [kPa],
Δ – slope vapor pressure curve [kPa °C^{-1}],
γ – psychrometric constant [kPa °C^{-1}],
The PM model uses a hypothetical green grass reference surface that is actively growing and is adequately watered with an assumed height of 0.12m, with a surface resistance of 70s m^{-1} and an albedo of 0.23 (Allen et al., 1998) which closely resemble evapotranspiration from an extensive surface of green grass cover of uniform height, completely shading the ground

and with no water shortage. This methodology is generally considered as the most reliable, in a wide range of climates and locations, because it is based on physical principles and considers the main climatic factors, which affect evapotranspiration.

Need for reduced-set methods

The main limitation to generalized application of this methodology in irrigation practice is the time and cost involved in daily acquisition and processing of the necessary meteorological data. Additionally, the number of meteorological stations where all these parameters are observed is limited, in many areas of the globe. The number of stations where *reliable* data for these parameters exist is an even smaller subset.

There are also concerns about the accuracy of the observed meteorological parameters (Droogers and Allen, 2002), since the actual instruments, specifically pyranometers (solar radiation) and hygrometers (relative humidity), are often subject to stability errors. It is common to see a drift, of as much as 10 percent, in pyranometers (Samani, 2000, 1998). Henggeler et al. (1996) have observed that hygrometers loose about 1 percent in accuracy per installed month. There are also issues related to the proper irrigation and maintenance of the reference grass, at the weather stations. Jensen et al. (1997) observed that many weather stations are often not irrigated or inadequately irrigated, during the summer months, and thus the use of relative humidity and air temperature from these stations could introduce a bias in the computed values for *ETo*. Additionally, they observed that the measured values of solar radiation, *Rs*, are not always reliable or available and that wind data are quite site specific, unavailable, or of questionable reliability. Thus, they recommend the use of *ETo* equations that require fewer variables. These authors compared various methods, including FAO Penman Monteith, PM, and Hargreaves and Samani, HS, with lysimeter data and noted r^2 values of 0.94-0.97, with monthly SEE values of 0.30-0.34mm. Based on these data they concluded that the differences in *ETo* values, calculated by the different methods, are minor when compared with the uncertainties in estimating actual crop evapotranspiration from ETo. Additionally, these equations can be more easily used in adaptive or smart irrigation controllers that adjust the application depth according to the daily *ETo* demand (Shahidian et al., 2009).

This has created interest and has encouraged development of practical methods, based on a single or a reduced number of weather parameters for computing *ETo*. These models are usually classified according to the weather parameters that play the dominant role in the model. Generally these classifications include the *temperature-based models* such as Thornthwaite (1948); Blaney-Criddle (1950) and Hargreaves and Samani (1982); The *radiation models* which are based on solar radiation, such as Priestly-Taylor (1972) and Makkink (1957); and the *combination models* which are based on the energy balance and mass transfer principles and include the Penman (1948), modified Penman (Doorenbos and Pruitt, 1977) and FAO PM (Allen et al., 1998).

Objectives and methods

The objective of this chapter is to review the underlying principles and the genesis of these methodologies and provide some insight into their applicability in various climates and regions. To obtain a global view of the applicability of the reduced-set equations, each equation is presented together with a review of the published studies on its regional calibration as well as its application under different climates.

The main approach for evaluation and calibration of the reduced-set equations has been to use the PM methodology or lysimeter measurements as the benchmark for assessing their performance. Usually a linear regression equation, established with PM *ETo* values or lysimeter readings plotted as the dependent variable and values from the reduced-set equation plotted as the independent variable. The intercept, a, and calibration slope, b, of the best fit regression line, are then used as regional calibration coefficients:

$$ET_o PM = a + b(ET_o Equation) \qquad (2)$$

The quality of the fit between the two methodologies is usually presented in terms of the coefficient of determination, r^2, which is the ratio of the explained variance to the total variance or through the Root Mean Square Error, *RMSE*:

$$RMSE = \sqrt{\frac{1}{n}\sum_{i=1}^{n}\left(ETo_{yi} - ETo_{PM}\right)^2} \qquad (3)$$

and the mean Bias error:

$$MBE = \frac{1}{n}\sum_{i=1}^{n}\left(ETo_{yi} - ETo_{PM}\right) \qquad (4)$$

where n is the number of estimates and ETo_{yi} is the estimated values from the reduced-set equation.

2. Temperature based equations

Temperature is probably the easiest, most widely available and most reliable climate parameter. The assumption that temperature is an indicator of the evaporative power of the atmosphere is the basis for temperature-based methods, such as the Hargreaves-Samani. These methods are useful when there are no data on the other meteorological parameters. However, some authors (McKenny and Rosenberg, 1993, Jabloun and Sahli, 2007) consider that the obtained estimates are generally less reliable than those which also take into account other climatic factors.

Mohan and Araumugam (1995) and Nandagiri and Kovoor (2006) carried out a multivariate analysis of the importance of various meteorological parameters in evapotranspiration. They concluded that temperature related variables are the most crucial required inputs for obtaining ETo estimates, comparable to those from the PM method across all types of climates. However, while wind speed is considered to be an important variable in arid climate, the number of sunshine hours is considered to be the more dominant variable in sub-humid and humid climates.

2.1 The Hargreaves- Samani methodology

Hargreaves, using grass evapotranspiration data from a precision lysimeter and weather data from Davis, California, over a period of eight years, observed, through regressions, that for five-day time steps, 94% of the variance in measured *ET* can be explained through average temperature and global solar radiation, *Rs*. As a result, in 1975, he published an equation for predicting *ETo* based only on these two parameters:

$$ET_o = 0.0135\ R_s(T + 17.8) \tag{5}$$

where Rs is in units of water evaporation, in mm day^{-1}, and T in °C. Subsequent attempts to use wind velocity, U_2, and relative humidity, RH, to improve the results were not encouraging so these parameters have been left out (Hargreaves and Allen, 2003).

The clearness index, or the fraction of the extraterrestrial radiation that actually passes through the clouds and reaches the earth's surface, is the main energy source for evapotranspiration, and later studies by Hargreaves and Samani (1982) show that it can be estimated by the difference between the maximum, T_{max}, and the minimum, T_{min} daily temperatures. Under clear skies the atmosphere is transparent to incoming solar radiation so the T_{max} is high, while night temperatures are low due to the outgoing longwave radiation (Allen et al., 1998). On the other hand, under cloudy conditions, T_{max} is lower, since part of the incoming solar radiation never reaches the earth, while night temperatures are relatively higher, as the clouds limit heat loss by outgoing longwave radiation. Based on this principle, Hargreaves and Samani (1982) recommended a simple equation to estimate solar radiation using the temperature difference, ΔT:

$$\frac{R_s}{R_a} = K_T\ (T_{max} - T_{min})^{0.5} \tag{6}$$

where Ra is the extraterrestial radiation in mm day^{-1}, and can be obtained from tables (Samani, 2000) or calculated (Allen et al., 1998). The empirical coefficient, K_T was initially fixed at 0.17 for Salt Lake City and other semi-arid regions, and later Hargreaves (1994) recommended the use of 0.162 for interior regions where land mass dominates, and 0.190 for coastal regions, where air masses are influenced by a nearby water body. It can be assumed that this equation accounts for the effect of cloudiness and humidity on the solar radiation at a location (Samani, 2000). The clearness index (Rs/Ra) ranges from 0.75 on a clear day to 0.25 on a day with dense clouds.

Based on equations (5) and (6), Hargreaves and Samani (1985) developed a simplified equation requiring only temperature, day of year and latitude for calculating ETo:

$$ET_o = 0.0135\ K_T\ (T + 17.78)(T_{mzx} - T_{min})^{0.5} R_a \tag{7}$$

Since K_T usually assumes the value of 0.17, sometimes the 0.0135 K_T coefficient is replaced by 0.0023. The equation can also be used with Ra in MJ m^{-2} day^{-1}, by multiplying the right hand side by 0.408.

This method (designated as HS in this chapter) has produced good results, because at least 80 percent of ETo can be explained by temperature and solar radiation (Jensen, 1985) and ΔT is related to humidity and cloudiness (Samani and Pessarakli, 1986). Thus, although this equation only needs a daily measurement of maximum and minimum temperatures, and is presented here as a temperature-based method, it effectively incorporates measurement of radiation, albeit indirectly. As will be seen later, the ability of the methodology to account for both temperature and radiation provides it with great resilience in diverse climates around the world.

Sepashkhah and Razzaghi (2009) used lysimeters to compare the Thornthwaithe and the HS in semi-arid regions of Iran and concluded that a calibrated HS method was the most accurate method. Jensen et al.(1997) compared this and other ETo calculation methods and concluded that the differences in ETo values computed by the different methods are not larger than those introduced as a result of measuring and recording weather variables or the uncertainties associated with estimating crop evapotranspiration from *ETo*. López-Urrea et al. (2006) compared seven *ETo* equations in arid southern Spain with Lysimeter data, and observed daily RMSE values between 0.67 for FAO PM and 2.39 for FAO Blaney-Criddle. They also observed that the Hargreaves equation was the second best after PM, with an RMSE of only 0.88.

Since the HS method was originally calibrated for the semi-arid conditions of California, and does not explicitly account for relative humidity, it has been observed that it can overestimate *ETo* in humid regions such as Southeastern US (Lu et al. 2005), North Carolina (Amatya et al. 1995), or Serbia (Trajkovic, 2007).

In Brasil, Reis et al. (2007) studied three regions of the Espírito Santo State: The north with a moderately humid climate, the south with a sub-humid climate, and the mountains with a humid climate (Table 1). The HS equation overestimated *ETo* in all three regions by as much as 32%, but the performance of the HS equation improved progressively as the climate became drier. Only further south, at a latitude of 24° S, and in a warm temperate climate did HS provide good agreement with PM, though still with a small overestimation. Borges and Mendiondo (2007) obtained an r^2 of 0.997 for HS when compared to PM, when using a calibrated α of 0.0022 (Sept-April) and 0.0020 for the rest of the year.

On the other hand, in dry regions such as Mahshad, Iran and Jodhpur, India, the HS equation tends to underestimate *ETo* by as much as 24% (Rahimkoob, 2008; Nandagiri and Kovoor, 2006). Rahimkoob (2008) studied the *ETo* estimates obtained from the HS equation in the very dry south of Iran. His data indicate that the HS equation fails to calculate *ETo* values above 9 mm day^{-1}, even when the PM reaches values of more than 13 mm day^{-1} (Fig. 1).

Wind removes saturated air from the boundary layer and thus increases evapotranspiration (Brutsaert, 1991). Since most of the reduced-set equations do not explicitly account for wind speed, it is natural for the calibration slope to be influenced by this parameter. Itensifu et al. (2003) carried out a major study using weather data from 49 diverse sites in the United States. They obtained ratios ranging from 0.805 to 1.242 between HS and PM and concluded that the HS equation has difficulty in accounting for the effects of high winds and high vapor pressure deficits, typical of the Great Plains region. They also observed that the HS equation tends to overestimate *ETo* when mean daily *ETo* is relatively low, as in most sites in the eastern region of the US, and to underestimate when *ETo* is relatively high, as in the lower Midwest of the US. As will be seen later, this seems to be a common issue with most of the reduced set evapotranspiration equations (see section 4.3, Fig. 7).

For the Mkoji sub-catchment of the Great Ruaha River in Tanzania, Igbadun et al. (2006) calculated the monthly *ETo* values of three very distinct areas of the catchment: the humid Upper Mkoji with an altitude of 1700m, the middle Mkoji with an average altitude of 1100 m, and the semi-arid lower Mkoji with an altitude of 900m. Their data indicate a strong relation between the monthly average wind speed and the performance of the HS equation as measured by the slope of the calibration equation (PM/HS ratio). Although the three areas have distinct climates, the HS equation clearly underestimated ETo for wind speed values below 2-2.3 ms^{-1}, and overestimated it for higher wind speed values (Fig. 2).

Trajkovic, et al. (2005) studied the HS equation in seven locations in continental Europe with different altitudes (42-433m) with RH ranging from 55 to 71%, representative of the distinct climates of Serbia. Their data show that despite the different altitudes and climatic conditions, wind speed was the major determinant for the calibration of the HS equation (Fig. 3). The results from these works indicate that wind is the main factor affecting the calibration of the HS equation and that the equation should be calibrated in areas with very high or low wind speeds.

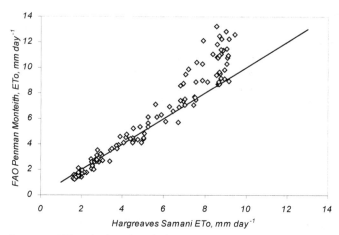

Fig. 1. Relation between *ETo* calculated with the HS equation and the PM for the dry conditions of Abadan, Iran. The Hargreaves Samani equation fails to calculate *ETo* values above 9 mm day⁻¹ (data kindly provided by Rahimkoob)

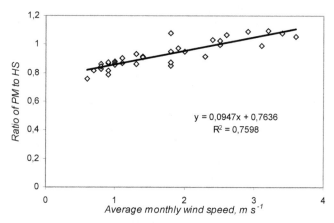

Fig. 2. Correlation between average wind speed and the calibration slope in distinct climates of the Great Ruana River in Tanzania (based on the original data from Igbadun et al. 2006).

Jabloun and Sahli (2008) studied eight stations in the semi-arid Tunisia and concluded that in inland stations, HS tends to overestimate *ETo* due to high ΔT values. In the coastal station of Tunis, HS underestimated *ETo* values, which they attributed to an underestimation of *Rs*. Various attempts have been made to improve the accuracy of the HS equation through incorporation of additional measured parameters, such as rainfall (Droogers and Allen, 2002) and altitude (Allen, 1995). These methodologies have had limited global application, probably because *ETo* is influenced by a combination of different parameters, and although in a certain region there appears to be a good correlation between the calibration slope and a certain parameter, this might not be so in a different climate.

The alternative is to use regional calibration, in which, based on the climatic characteristics of the region, the *ETo* calculated by the HS equation is adjusted to account for the combined effect of the dominant climate parameters, and thus accuracy of the equations is improved (Teixeira et al., 2008). Table 1 presents a compilation of most of the published studies on the regional calibration of the HS equation. This compilation contains 33 published works covering 21 countries with all types of climatic conditions according to the Koppen classification. Whenever various stations from a similar climate were studied, only parameters from one representative station are presented. In some studies, HS and PM were calibrated against a third methodology (such as Pan A) and thus no direct calibration parameters for the PM/HS regression were provided. In these cases, a linear regression was obtained by plotting the PM calibration equation as the dependent variable and the HS calibration equation as the independent variable. The parameters of the resulting regression equation are then presented as the PM-HS calibration parameters.

In order to contextualize the information and allow for extension of the results to other regions with a similar climate, the locations are grouped according to Koppen climate classification. These calibration coefficients can be used in the area where they were obtained or can be extrapolated for areas with similar conditions where no actual calibration has been carried out yet.

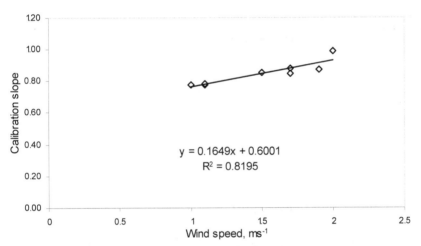

Fig. 3. Correlation between wind speed and the calibration slope for seven different locations in Serbia, representing the diverse local climates (original data from Trajkovic, 2005).

2.2 The Thornthwaite method

Thornthwaite (1948) devised a methodology to estimate *ETo* for short vegetation with an adequate water supply in certain parts of the USA. The procedure uses the mean air temperature and number of hours of daylight, and is thus classified as a temperature based method. Monthly *ETo* can be estimated according to Thornthwaite (1948) by the following equation:

$$Et_0 = ET_0 sc\left(N/12\right)\left(dm/30\right) \tag{8}$$

Country	Station	latitude m	Altitude m	Köppen classification	Rainfall mm	RH %	U2 ms⁻¹	intercept a	slope b	R2	RMSE	Source
Arid												
Desert												
China, NW	Shandan Heihe R.	38°90' N	1483	BWk	250	40	1.98	0.5431	1.148			Zhao et al. 2005
China, NW	Minle	38°80' N	2271	BWk	100	35.3	3.2	-0.32	1.065			Zhao et al. 2005
US	Aquila	33°56' N	655	BWh	195		3.2	0.0378	1.3155			Alexandris, 2006
Steppe												
India	Jodhpur	26°18' N	224	BSh	402	38.9	2.1	-0.3827	1.1924			Nandagiri and Kovoor, 2006
India	Heydarabad	17°32' N	545	BSh	820	65.6	2.8	-1.97	1.48			Nandagiri and Kovoor, 2006
Syria	Tel Hadya	36°01' N	293	BSh	231	57.4	2.82		1.04			Stockle, 2004
Iran	Shiraz	30°07' N	1650	BSh	306	36.4	2.49	0.41	0.82	0.91		Razzaghi and Sepahskah, 2009
Iran	Shiraz	30°07' N	1650	BSh	305.6	36.4	2.49		1.13			Sepashkah and Razzaghi, 2009
México	Progreso (Yucatán)	21°17' N	2	BSh	511			-0.26	1.012	0.78		Bautista et al 2009
Dry summer												
Spain	Daroca (NE Spain)	41°07' N	779	Bsk	364	66.5	1.08	-0.203	0.93			Mártinez-Cob and Tejero-Juste, 2004
Spain	Zaragoza (NE Spain)	41°43' N	225	Bsk	353	73.7	2.43	-0.012	0.99			Mártinez-Cob and Tejero-Juste, 2004
Spain	Cordoba, inland	37°52' N	117	Bsk	696	63.3	1.6		1.06			Gavilán et al, 2008
Bolivia	Patacamaya and Oruro	17°15'S	3749	Bsk	375	57.4	1.2	0.8622	0.6422			Garcia et al 2004
Spain	Albacete	39°14' N	695	Bsk	283	68.7	1.08	0.34*	1.14*			Lopéz-Urra et al 2005
Spain	Cordoba, inland	37°51' N	110	Bsk	696	63.3	1.6	-1.49	1.3			Berengena and Gavilan, 2005
Tanzania	Lower Mkoji	7°80'	900	Bsh	520			-0.0027	0.9092			Igbadun et al
Mesothermal												
Mediterranean												
Spain	Malaga (Andalucia) Coast	36°40' N	7	Csa	531	68.1	1.9		0.962			Vanderlinden et al., 2004
Spain	Sevilla (Andalucia) interior	37°125' N	31	Csa	473	67.8	0.93		1.165			Vanderlinden et al., 2004
Portugal, S	La Mojonera, coast	37°45' N	142	Csa	272	62.3	1.9		1.27			Gavilán et al, 2008
Portugal	Evora	38°55' N	246	Csa	627	63.3	4.3		0.866			Santos and Maia, 2007
US	Davis	38°32' N	18.3	Csa	458	63.3	2.62		1.245			Alexandris, 2006
Portugal	Elvas	38°60' N	202	Csa	508	58.2	1.97	-0.844	1.04			Teixeira et al. 2008
Spain	Niebla (Andalucia)	37°21' N	52	Csa	702	65.3	1.3	-0.08	1.035	0.93		Gavilán et al, 2008
Spain	Vejer Frontera (Andalucia)	38° 17' N	24	Csa	571	69.4	2.9		1.404			Gavilán et al, 2008
Greece	Athens	38°23' N	100	Csa	371	61.8	1.87	0.264	0.781			Alexandris, 2006
USA	Prosser, WA	46°15' N	380	Csb	994	69.7	1.62		1.02	0.98		Stockle, 2004
Spain	Lleida	41°42' N	221	Csb	601	68.8	0.97		1.1	0.95		Stockle, 2004
Dry winter												
Tanzania	Middle Mkoji	8°30'	1070	Cwa	800			-0.4	0.955			Igbadun et al, 2006
Brasil	Douradas, Mato G. Sul	22°16'S	452	Cwa	1603	73.8	1.74	1.73	0.67	0.7		Fietz, 2004
Brasil	S. Mantiqueira, MG		1500	Cwb	2150			0.153	1.16			Pereira et al. 2009
fully humid												
Netherlands	Haarweg	51°58' N	9	Cfb	778	87.3	2.41		1.02	0.91		Stockle, 2004
US	Louisiana, inland	31° N	low land	Cfa	1500	92	0.82	-0.28	1.05			Fontenot, 2004
US	Louisiana, coastal	29° N	low land	Cfa	1500	88.7	0.6	-0.17	0.87			Fontenot, 2004
US	North Carolina, Plymouth	35°52'	6	Cfa	1299	80.2	4.9	0.03	0.83			Amatya et al. 1995
Brasil	Palotina, Paraná	24°18'S	310	Cfa	1700	73.8	1.74	-108	1		1.23	Syperreck, 2006
Brasil	Jacupiranga river, SP	24°29'S	52	Cfa	1879	91.5	0.97	-0.365	1.042			Borges and Mendiondo, 2007

Values in grey are annual averages obtained from Climwat data base.
When calibration parameters of the HS vs FAO PM were not directly provided, linear regression equations were established with FAO-56 PM daily ET0 estimates as the dependent variable and daily ET0 values estimated by HS as an independent variable. The parameters of the regression equation were then presented as the calibration parameters.

Country	Station	latitude	Altitude m	Classification Köppen	Rainfall mm	RH %	U_2 ms^{-1}	Regression adjustment intercept a	Regression adjustment slope b	R2	RMSE	Source
Microthermal												
Fully humid												
Serbia	Kragujevac	44°00' N	190	Dfa		75%	1		0.78		0.451	Trajkovic, 2005
Serbia	Belgrade	44°45' N	132	Dfa	684	69%	1.7		0.99			Trajkovic, 2005
Cro., Ser. Bos.	Zagreb, Sarajevo, etc.	42.6- 46.1	42-630	Dfb		68-76	1.0-1.9		0.424 [3]			Trajkovic, 2007
Canada	Southern Ontario, Drumbo	43°16' N	310	Dfb		79%	1.5		0.74	0.7	0.704	Sentelhas et al. 2010
Canada	Southern Ontario, Harrow	42°12' N	190	Dfb		73%	2.2		0.94	0.64	0.704	Sentelhas et al. 2010
Dry winter												
China	Tibete plateau- Yushu	33°06' N	3681	Dwb	200	45.4	0.83	0.347	0.883	0.91	0.622	Ye et al. 2009
Polar												
Bulgaria	Trace plain, Plovdiv	42°25' N	160	ET	492	73			1.11			Popova et al, 2006
Switzerland	Changins	46°24N	416	ET	904		2.5	-0.31	1.12	0.99		Xu and Singh, 2002
Tropical												
Winter dry												
México	Mérida (Yucatán)	20°56' N	15	Aw	11.74	77.5	1.23	0.1754	1.021	0.78		Bautista et al 2009
Tanzania	Upper Mkoji	9°00'	1700	Aw	1070			0.006	0.987			Igbadun et al
India	Kharagpur	13°00' N	921	Aw	940	66	1.9	-2.64	1.561			Kashyap and Panda, 2001
India	Bangalore		62	Aw	1506	92	2.12	-0.1063	1.0244			Nandagiri and Kovoor, 2006
Nigeria	Abeokuta	7°10' S	63	Aw	1506	92	2.12	-1.41	0.938			Adeboye, 2009
Nigeria	Abeokuta	7°10' S	823	Aw	1506	87.9	0.82	0.0025 [1]	16.8 [2]			Adeboye, 2009
Brasil	Goiânia, GO	16°28' S	75	Aw	1785	75.9	3.34	0.6923	0.3811	0.47		Oliveira et al. 2005
Brasil	Sooretama (South Espírito Santo)	19°22'S		Am				-2.62	1.572			Reis et al, 2007
Summer dry												
Fully humid												
Brasil	Campina Grande	7°14'S	550	As'	700	80	1.38	-0.488	0.893			Henrique, 2006
Fully humid												
Brasil	North Rio de Janeiro	21°19'S	13	Af	1172.9	73.1	0.3	-0.76	1			Mendonça et al 2003
Philipines	Los Banos	14°13' N	41	Af	1987	83.3	1.35		0.96	0.65		Stockle, 2004

* compared with lysimeter values

Values in grey are annual averages obtained from Climwat data base.

(1) (2) (3) Respectively, the KT, d and e of regionally calibrated HS equation, according to Equation 7 in the text.

When calibration parameters of the HS vsFAO PM were not directly provided, linear regression equations were established with FAO-56 PM daily ET0 estimates as the dependent variable and daily ET0 values estimated by HS as an independent variable. The parameters of the regression equation were then presented as the calibration parameters.

Table 1. Regional calibration for the Hargreaves Samani equation compiled from published works

Where N is the maximum number of sunny hours as a function of the month and latitude and dm is the number of days per month. ETo_{sc} is the gross evapotranspiration (without corrections) and can be calculated as:

$$Et_0sc = 16\left(\frac{10T_a}{I}\right)a \tag{9}$$

where T_a is the mean daily temperature (°C), a is an exponent as a function of the annual index: $a = 0.49239 + 1792 \times 10^{-5}I - 771 \times 10^{-7}I^2 + 675 \times 10^{-9}I^3$; and I is the annual heat index obtained form the monthly heat indecies:

$$I = \sum_{m=1}^{12}\left(\frac{T_m}{5}\right)1.514 \tag{10}$$

Bautista et al. (2009) found that the precision of the Thorntwaite methodology improved during the winter months in Mexico. Garcia et al. (2004) observed that under the dry and arid conditions of the Bolivian highlands the Thornthwaite equation strongly underestimates ETo because the equation does not consider the saturation deficit of the air (Stanhill, 1961; Pruitt, 1964; Pruitt and Doorenbos, 1977). Additionally, at high altitudes, the Thornthwaite equation also underestimates the effect of radiation, because the equation is calibrated for temperate low altitude climates. Studies in Brazil have shown that the underestimation of ETo produced by temperature-based equations under arid conditions, may be reduced by using the daily thermal amplitude instead of the mean temperature (Paes de Camargo, 2000) as in the case of the Hargreaves–Samani equation.

Gonzalez et al. (2009) studied the Thorthwaite method in the Bolivian Amazon. They observed that the Thornthwaite method underestimates evapotranspiration at all the three stations studied. This is expected, considering that normally this method leads to underestimations in humid areas (Jensen et al., 1990).

2.3 Blaney-Criddle method

The FAO Temperature Methodology recommended by Doorenbos and Pruitt (1977) is based on the Blaney-Criddle method (Blaney and Criddle, 1950), introducing a correction factor based on estimates of humidity, sunshine and wind.

$$ET_o = \alpha + \beta\left[p(0.46T + 8.13)\right] \tag{11}$$

where α and β are calibration parameters and p is the mean annual percentage of daytime hours. Values for α can be calculated using the daily RH_{min} and n/N as follows:

$$\alpha = 0.043RH_{min} - \left(\frac{n}{N}\right) - 1.41 \tag{12}$$

$$\frac{n}{N} = 2(Rs/Ra) - 0.5 \tag{13}$$

For windy South Nebraska, Irmak et al. (2008) compared 12 different ET methodologies and found that the Blaney–Criddle method was the best temperature method and it had an RMSE value (0.64 mm d⁻¹) which was similar to some of the combination methods. The obtained estimates were good and were within 3% of the ASCE-PM *ETo* with a high r² of 0.94. The estimates were consistent with no large under or over estimations for the majority of the dataset. They attributed this to the fact that, unlike most of the other temperature methods, this method takes into account humidity and wind speed in addition to air temperature.

Lee et al. (2004) compared various *ETo* calculation methods in the West Coast of Malaysia and concluded that the Blaney-Criddle method was the best, among the reduced-set equations, for estimating ET in the region. They also observed that HS gave the highest estimates followed by the Priestly-Taylor equation. Similarly, in the humid Goiânia region of Brazil, Oliveira et al. (2005) observed that the Blaney-Criddle method produced the best results, next to the full PM equation.

Various studies indicate that the Blaney-Criddle equation might show some bias under arid conditions. For semi-arid conditions of Iran, Dehghani Sanij et al. (2004) found the Blaney-Criddle and the Makkink method to overestimate ETo during the growing season. Lopéz-Urrea et al. (2006) compared seven different methods for calculating ETo in the semiarid regions of Spain and observed that the Blaney-Criddle method significantly over-estimated average daily ETo.

For arid conditions of Iran, Fard et al. (2009) compared nine different methodologies with lysimeter data and observed that the Turc and the Blaney-Criddle methods showed very close agreement with the lysimeter data, while PM showed moderate agreement with the lysimeter data. The other methods showed bias, systematically over estimating the lysimeter data (Fig. 4).

Although recognizing the historical value of the Blaney-Criddle method and its validity, the FAO Expert Commission on Revision of FAO Methodologies for Crop Water Requirements (Smith et al. 1992) did not recommend the method further, in view of difficulties in estimating humidity, sunshine and wind parameters in remote areas. Nevertheless, they emphasized the value of the method for areas having only the mean daily temperature, and where appropriate correction factors can be found.

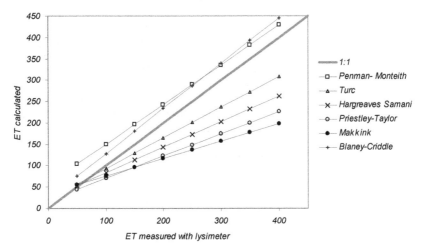

Fig. 4. Comparision of six ET methods with lysimeter data for Isfahan (adapted from Fard et al., 2009).

2.4 Reduced-set PM

The PM methodology has provisions for application in data-short situations (Allen et al. 1998), including the use of temperature data alone. The reduced-set PM equation requiring only the measured maximum and minimum temperatures uses estimates of solar radiation, relative humidity, and wind speed. Solar radiation, Rs, MJ m^{-2} d^{-1} can be estimated using equation 3 (Hargreaves and Samani, 1985) or using averages from nearby stations. For island locations Rs can be estimated as (Allen et al. 1998):

$$R_s = 0.7R_a - b \qquad (14)$$

where b is an empirical constant with a value of 4 MJ m^{-2} d^{-1} . Relative humidity can be estimated by assuming that the dewpoint temperature is approximately equal to T_{min} (Allen 1996; Allen et al. 1998) which is usually experienced at sunrise. In this case, e_a can be calculated as:

$$e_a = e^o(T_{min}) = 0.611 \exp\left[\frac{17.27T_{min}}{T_{min} + 237.3}\right] \qquad (15)$$

where $e^o(T_{min})$ is the vapour pressure at the minimum temperature, expressed in mbar. For wind speed, Allen et al. (1998) recommend using average wind speed data from nearby locations or using a wind speed of 2 m s^{-1}, since, they consider, the impact of wind speed on the ETo results is relatively small, except in arid and windy areas. The soil heat flux density, G, for monthly periods can be estimated as:

$$G_i = 0.07(T_{i+1} - T_{i-1}) \qquad (16)$$

where G_i is the soil heat flux density in month I in MJ m^{-2} d^{-1}; and T_{i+1} and T_{i-1} are the mean air temperatures in the previous and following months, respectively.

Allen (1995) evaluated the reduced-set PM (using only $Tmax$ and $Tmin$) and HS using the mean annual monthly data from the 3,000 stations in the FAO CLIMWAT data base, with the full PM serving as the comparative basis. He found little difference in the mean monthly ETo between the two methods. Wright et al. (2000) found similar results in Kimberly, and 75 years of data from California (Hargreaves and Allen, 2003). Other data generally indicate that the reduced-set PM performs better in humid areas (Popova, 2005, Pereira et al., 2003), while HS performs better in dry climates (Temesgen et al. 2005, Jabloun et al. 2008).

Trajkovic (2005) compared the reduced-set PM, Hargreaves, and Thornthwaite temperature-based methods with the full PM in Serbia and found that the reduced-set PM estimates were better than those produced from the Hargreaves and Thornthwaite equations. Popova et al. (2006) found the reduced-set PM to provide more accurate results compared to the Hargreaves equation, which tended to overestimate reference evapotranspiration in the Trace plain in south Bulgaria. Jabloun and Sahli (2008) also found the Hargreaves equation to overestimate reference evapotranspiration in Tunisia and found the reduced-set PM equation to provide better estimates. Nevertheless, the reduced-set PM can produce poor results in areas where wind speed is significantly different from 2 ms^{-1} (Trajkovic, 2005).

3. Radiation based methods

It is known that water loss from a crop is related to the incident solar energy, and thus it is possible to develop a simple model that relates solar radiation to evapotranspiration.

Various models have been developed, over the years, for relating the measured net global radiation to the estimated reference evapotranspiration; such as the Priestley-Taylor method (1972), the Makkink method (1957), the Turc radiation method (1961), and the Jensen and Haise method (1965).

Irmak et al. (2008) compared 11 ET models and studied the relevance of their complexity for direct prediction of hourly, daily and seasonal scales. They concluded that radiation is the dominant driver of evaporative losses, over seasonal time scales, and that other meteorological variables, such as temperature and wind speed, gained importance in daily and hourly calculations.

3.1 The Priestley-Taylor method

The Priestley-Taylor method (Priestley and Taylor, 1972; De Bruin, 1983) is a simplified form of the Penman equation, that only needs net radiation and temperature to calculate ETo. This simplification is based on the fact that ETo is more dependant on radiation than on relative humidity and wind. The Priestly-Taylor method is basically the radiation driven part of the Penman Equation, multiplied by a coefficient, and can be expressed as:

$$ET_o = \alpha \frac{\Delta(R_n - G)}{\Delta + \gamma} + \beta \tag{17}$$

where α and β are calibration factors, assuming values of 1.26 and 0, respectively. This model was calibrated for Switzerland (Xu and Singh, 1998) and values of 0.98 and 0.94 were obtained for α and β, respectively. In the Priestley-Taylor equation, evapotranspiration is proportional to net radiation, while in the Makkink equation (section 3.2), it is proportional to short-wave radiation.

Van Kraalingen and Stol (1997) found that application of the Priestly-Taylor equation during the Dutch winter months was not possible because it is based on net radiation. Since net radiation is often negative in the winter, it predicts dew formation, whereas the actual ET is positive. The situation would be different for a humid climate such as the Philippines, or in a semi-arid climate such as Israel, where the equation should compare well with PM.

Irmak et al. (2003) calibrated the Priestly-Taylor method against the FAO PM method using 15 years of climate data (1980–1994) in humid Florida, United States. The monthly values of the calibration coefficient (Fig. 5) show a considerable seasonal variation, aside from the natural difference in annual values. In general, the calibration coefficients are lower in winter months indicating that the Priestley and Taylor method underestimates ETo, and they are higher than 1.0 during the summer months, indicating that the method overestimates during the summer months. The long-term average lowest calibration values were obtained in January and December (0.70) and the highest values in July (1.10). These results indicate the importance of developing monthly calibration coefficients for regional use based on historic records. For the semi-arid conditions of southern Portugal, the authors also observed that the Priestley-Taylor method over-estimates daily ETo during the summer months (Shahidian et al., 2007).

Shuttleworth and Calder (1979) showed that Priestley-Taylor significantly underestimates wet forest evaporation, but also overestimates dry forest transpiration by as much as 20%. Berengena and Gavilán (2005) found that the Priestley–Taylor equation shows a considerable tendency to underestimate ETo, on average 23%, under convective conditions.

They concluded that the Priestly-Taylor equation is very sensitive to advection, and local calibration does not ensure an acceptable level of accuracy.

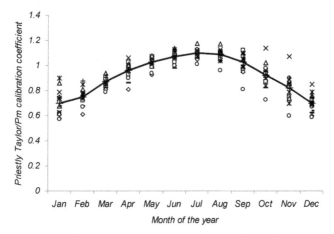

Fig. 5. Average monthly calibration coefficient for the Priestly-Taylor equation against PM for humid southern United States (based on data from Irmak et al. 2003).

3.2 The Makkink method

The Makkink method can be seen as a simplified form of the Priestley-Taylor method and was developed for grass lands in Holland. The difference is that the Makkink method uses incoming short-wave radiation Rs and temperature, instead of using net radiation, Rn, and temperature. This is possible, because on average, there is a constant ratio of 50% between net radiation and short wave radiation. The equation can be expressed as:

$$Et_o = \alpha \frac{\Delta}{\Delta + \gamma} \frac{R_s}{2,45} + \beta \qquad (18)$$

where α is usually 0.61, and β is -0.012. Doorenbos and Pruitt (1975) proposed the FAO Radiation method based on the Makkink equation (1957), introducing a correction factor based on estimates for wind and humidity conditions to compensate for advective conditions. This radiation method has been proven valid, in particular under humid conditions, but can differ systematically from the PM reference method under special conditions, such as during dry months (Bruin and Lablands, 1998).

It has also been observed that it is difficult to use this radiation based method during winter months: Van Kraalingen and Stol (1997) found that application of the Makkink equation in Dutch winter months was not possible, though the Makkink equation did not produce negative values for ET, as was the case with the Priestley-Taylor method. Bruin and Lablans (1998) also concluded that there is no relationship between Makkink and PM in the winter months, December and January, since Makkink's method has no physical meaning, in this period.

It is reasonable to expect the Makkink and the Priestley-Taylor equations to compare well with the Penman's method, since in all these approaches the radiation terms are dominant and radiation is the main driving force for evaporation in short vegetation.

ET models tend to perform best in climates in which they were designed. A study by Amayta et al. (1995) showed that while the Makkink model generally performed well in North Carolina, the model underestimated *ETo* in the peak months of summer. Yet, the Makkink model shows excellent results in Western Europe where it was designed, both in comparison to PM as well as to the measured ETo data (Bruin and Lablans 1998, Xu and Singh 2000, Bruin and Stricker 2000, Barnett et al., 1998).

3.3 The Turc method

Also known as the Turc-Radiation equation, this method was presented by Turc in 1961, using data from the humid climate of Western Europe (France). This method only uses two parameters, average daily radiation and temperature and for RH>50% can be expressed as:

$$ET_p = \alpha\left((23,9001R_s) + 50\right)\left(\frac{T}{T+15}\right) \tag{19}$$

And for RH < 50% as:

$$ET_p = \alpha\left((23,9001R_s) + 50\right)\left(\frac{T}{T+15}\right)\left(1 + \left(\frac{50-RH}{70}\right)\right) \tag{20}$$

Where α is 0.01333 and *Rs* is expressed in MJ m^{-2} day^{-1}.

Yoder et al. (2005) compared six different ET equations in humid southeast United States, and found the Turc equation to be second best only to the full PM. Jensen et al. (1990) analyzed the properties of twenty different methods against carefully selected lysimeter data from eleven stations, located worldwide in different climates. They observed that the Turc method compared very favorably with combination methods at the humid lysimeter locations. The Turc method was ranked second when only humid locations were considered, with only the Penman-Monteith method performing better. Trajkovic and Stojnic (2007) compared the Turc method with full PM in 52 European sites and found a SEE (Standard Error of Estimate) of between 0.10 and 0.37 mm d^{-1}. They also found that the reliability of the Turc method depends on the wind speed (Fig. 6). The Turc method overestimated PM ETo in windless locations and generally underestimated ETo in windy locations.

Amatya et al. (1995) compared 5 different ETo methodologies in North Carolina and concluded that the Turc and the Priestley-Taylor methods were generally the best in estimating ETo. They observed that all other radiation methods and the temperature based Thorntwaite method underestimated the annual ET by as much as 16%.

Kashyap and Panda (2001) compared 10 different methods with lysimeter data in the sub humid Kharagupur region of India and observed that the Turc method had a deviation of only 2.72% from lysimeter values, followed by Blaney-Criddle with a 3.16% and Priestly Taylor with a 6.28% deviation (Fig. 7). The Kashyap and Panda data are also important because they show that under sub humid conditions, most of the equations, including the PM, tend to overestimate when evapotranspiration is low, and underestimate when it is high.

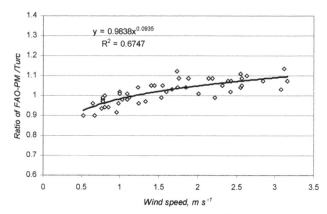

Fig. 6. Effect of wind on the ratio of evapotranspiration calculated with the FAO PM and the Turc methods (based on data from Trajkovic and Stojnic (2007), using average annual values).

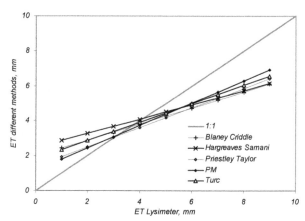

Fig. 7. Comparison of various ETo methods with Lysimeter readings in the sub-humid region of Kharagpur, India (adapted from Kashyap and Panda, 2001).

For Florida, Martinez and Thepadia (2010) compared the reduced-set PM equation with various temperature and radiation based equations and concluded that in the absence of regionally calibrated methods, the Turc equation has the least error and bias when using measured maximum and minimum temperatures. They also observed that the reduced-set PM and Hargreaves equations overestimate ET.

Fontenote (2004) studied the accuracy of seven evapotranspiraiton models for estimating grass reference ET in Louisiana. He observed that, statewide and in the coastal region, the Turc model was the most accurate daily model with a MAE of 0.26mm day[-1]. Inland, the Blaney-Criddle performed best with a MAE of 0.31mm day[-1] (Fig. 8).

Hence, it can be safely concluded that the Turc model can be expected to perform well in warm, humid climates such as those found in North Carolina (Amatya et al., 1995), India (George et al., 2002), and Florida (Irmak et al., 2003; Martinez and Thepadia, 2010).

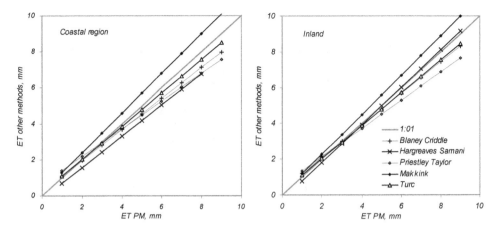

Fig. 8. Comparison of five ET methods with PM in two different regions of Louisiana (Adapted from Fontenote, 1999).

3.4 The Jensen and Haise method

This method was derived for the drier parts of the United States and is based on 3,000 observations of ET. Jensen and Haise used 35 years of measured evapotranspiration and solar radiation to derive the equation, based on the assumption that net radiation is more closely related to ET than other variables such as air temperature and humidity (Jensen and Haise, 1965). The equation can be expressed as:

$$ET = C_t (T - T_x) R_s \tag{21}$$

The original study of Jensen and Haise provides a calculation procedure to obtain Rs from the cloudiness, Cl, and the solar and sky radiation flux on cloudless days. The temperature Constant, C_t, and the intercept of the temperature exis, T_x, can be calculated as follows:

$$C_t = \frac{1}{\left[\left(45 - \dfrac{h}{137} \right) + \left(\dfrac{365}{e^0 (T_{max}) - e^0 (T_{min\,i})} \right) \right]} \tag{22}$$

and

$$T_x = -2.5 - 0.14 \left(e^0 (T_{max}) - e^0 (T_{min}) \right) - \frac{h}{500} \tag{23}$$

where h is the altitude of the location in m, Rs is solar radiation (MJ m^{-2} d^{-1}); $e^0 T_{max}$ and $e^0 T_{min}$ are vapour pressures of the month with the mean maximum temperature and the month with the mean minimum temperature, respectively, expressed in mbar.

For the humid and rainy Rio Grande watershed in Brazil, Pereira et al. (2009) compared 10 different equations and concluded that the methods based on solar radiation are more accurate than those based only on air temperature, with the Jensen and Haise method presenting the smallest MBE, and thus being the method most recommended for this region.

4. Conclusions

Both temperature and radiation can be used successfully to calculate daily *ETo* values with relative accuracy. All the equations can be used for areas that have a climate that is similar to the one for which the equations were originally developed; while most of the equations can be used with some confidence for areas with moderate conditions of humidity and wind speed.
Regional calibration, especially if including monthly calibration coefficients, is important in decreasing the bias of the ETo estimates. Wind speed can greatly influence the results obtained with reduced-set equations, since wind removes the boundary layer from the leaf surface and can significantly increase evapotranspiration. Relative Humidity is another important factor that can affect the results.
Globally, it is observed that the Turc equation is highly recommended for humid or semi-humid areas, where it can produce very good results even without calibration, while the Thornthwaite equation tends to underestimate *ETo*.
The Priestley-Taylor and the Makkinik equations should not be used in the winter months in locations with high latitude, such as northern Europe.
Both the Hargreaves and the reduced-set Panman-Monteith can be effectively used with only temperature measurements, although the results can be improved if wind speed is taken into consideration.
The use of the reduced-set equations can be very important in actual irrigation management, since the error involved in using these equations can be much smaller than that resulting from using data from a weather station located many miles away.

5. References

Allen RG (1993) Evaluation of a temperature difference method for computing grass reference evapotranspiration. Report submitted to the Water Resources Develop. and Man. Serv., Land and Water Develop. Div., FAO, Rome. 49 p.

Allen RG, Pereira LS, Raes D, Smith M (1998) Crop evapotranspiration: Guidelines for computing crop requirements. Irrigation and Drainage Paper No. 56, FAO, Rome, Italy.

Amatya DM, Skaggs RW, Gregory JD(1995) Comparison of Methods for Estimating REF-ET. Journal of Irrigation and Drainage Engineering 121:427-435.

Bautista F, Bautista D, Delgado-Carranza (2009) Calibration of the equations of Hargreaves and Thornthwaite to estimate the potential evapotranspiration in semi-arid and sub-humid tropical climates for regional applications. Atmósfera 22(4): 331-348

Berengena J, Gavilán P (2005) Reference evapotranspiration estimation in a highly advective semiarid environment. J. Irrig. Drain. Eng. ASCE 131 (2):147–163.

Borges AC, Mendiondo EM (2007) Comparação entre equações empíricas para estimativa da evapotranspiração de referência na Bacia do Rio Jacupiranga. Revista Brasileira de Engenharia Agrícola e Ambiental 11(3): 293–300.

Blaney, HF, Criddle, WD (1950). Determining water requirements in irrigated áreas from climatological and irrigation data. In ISDA Soil Conserv. Serv., SCS-TP-96,

Blaney HF, Criddle WD. (1962). Determining Consumptive Use and Irrigation Water Requirements. USDA Technical Bulletin 1275, US Department of Agriculture, Beltsvill

Bruin, HAR Lablans, WN, (1998) Reference crop evapotranspiration determined with a modified Makkink equation. Hydrological Processes

Brutsaert, W (1991) Evapotration into the atmosphere. D. Reidel Publishing Company.

DehghaniSanij H, Yamamoto T, Rasiah V (2004) Assessment of evapotranspiration estimation models for use in semi-arid environments. Agricultural Water Management 64: 91–106

Hargreaves, G.H. (1994). Simplified coefficients for estimating monthly solar radiation in North America and Europe. Departmental Paper, Dept. of Biol. And Irrig. Engrg., Utah State University, Logan, Utah

Doorenbos J., Pruit WO. (1977) Guidelines for predicting crop water requirements. FAO irrigation and drainagem paper, 24

Droogers P, Allen, RG (2002) Estimating reference evapotranspiration under inaccurate data conditions. *Irrig. Drain. Syst.* 16(1): 33-45.

Fontenot RI (2004) An evaluation of reference evapotranspiration models in Louisiana, MSc thesis, Louisiana State University, August.

Garcia M, Raes D, Allen R, Herbas C, (2004) Dynamics of reference evapotranspiration in the Bolivian highlands (Altiplano). Agric. Forest Meteorol. 125: 67–82.

George BA, Reddy BRS, Raghuvanshi NS, Wallender WW (2002) Decision support system for estimating reference evapotranspiration. J. Irrig. Drain. Eng. 128(1): 1–10.

Hargreaves GH, Samni ZA. (1982) Estimation of potential evapotranspiration. Journal of Irrigation and Drainage Division, Proceedings of the American Society of Civil Engineers 108: 223-230

Hargreaves GH, Samani ZA. (1985) Reference crop evapotranspiration from temperature. Appl Engine Agric. 1(2):96–99.

Hargreaves GH, Allen RG (2003) History and Evaluation of Hargreaves Evapotranspiration Equation. Journal of Irrigation and Drainage Engineering129(1): 53-63.

Henggeler J.C., Z. Samani, M.S. Flynn, J.W. Zeitler (1996) Evaluation of various evapotranspiration equations for Texas and New Mexico. Proceeding of Irrigation Association International Conference, San Antonio, Texas

Igbadun H, Mahoo H, Tarimo A, Salim B (2006) Performance of Two Temperature-Based Reference Evapotranspiration Models in the Mkoji Sub-Catchment in Tanzania. Agricultural Engineering International: the CIGR Ejournal. Manuscript LW 05 008. Vol. VIII. March,

Irmak S, Irmak A, Allen RG, Jones JW (2003) Solar and Net Radiation-Based Equations to Estimate Reference Evapotranspiration in Humid Climates. Journal of irrigation and drainage engineering. September/October

Irmak S, Istanbulluoglu, E, Irmak A. (2008) An Evaluation of Evapotranspiration model complexity against performance in comparison with Bowen Ration Energy Balance measurements. Transactions of the ASABE 51(4):1295-1310

Jabloun M, Sahli A (2007) Ajustement de l'e´quation de Hargreaves-Samani aux conditions climatiques de 23 stations climatologiques Tunisiennes. Bulletin Technique no. 2. Laboratoire de Bioclimatologie, Institut National Agronomique de Tunisie, 21 p.

Jabloun M, Sahli A (2008) Evaluation of FAO-56 methodology for estimating reference evapotranspiration using limited climatic data. Application to Tunisia. Agricultural Water Management 95: 707-715.

Jensen ME, Haise HR (1963) Estimating evapotranspiration from solar radiation. Journal of Irrigation and Drainage Division, Proc. Amer. Soc. Civil Eng. 89:15–41.

Jensen DT, Hargreaves GH, Temesgen B, Allen RG (1997) Computation of ETo under non ideal conditions. J. Irrig. Drain. Eng. ASCE 123 (5): 394–400.

Kashyap PS, Panda RK (2001) Evaluation of evapotranspiration estimation methods and development of crop-coefficients for potato crop in a sub-humid region. Agricultural Water Management 50: 9-25

López-Urrea R, Martín de Santa Olalla F, Fabeiro C, Moratalla A (2006) Testing evapotranspiration equations using lysimeter observations in a semiarid climate. Agricultural Water Management 85: 15–26.

Lu, J, Sun G, McNulty S, Amatya DM (2005) A Comparison of Six Potential Evapotranspiration Methods for Regional Use in the Southeastern United States. Journal of the American Water Resources Association (JAWRA) 41(3):621-633.

Makkink GF. (1957) Testing the Penman formula by means of lysimeters. Journal of the Institution of Water Engineers 11: 277-28

Martinez, A.M., Thepadia, M. (2010) Estimating Reference Evapotranspiration with Minimum Data in Florida. J. Irrig. Drain. Eng. 136(7): 494-501

McKenney, M. S. and N. J. Rosenberg, (1993) Sensitivity of some potential evapotranspiration estimation methods to climate change. - *Agricultural and Forest Meteorology* 64, 81-110.

Mohan S, Arumugam N (1996) Relative importance of meteorological variables in evapotranspiration: Factor analysis approach. Water Resour. Manage. 10, 1–20.

Nandagiri L, Kovoor GM (2006) Performance Evaluation of Reference Evapotranspiration Equations across a Range of Indian Climates. Journal of Irrigation and Drainage Engineering 132(3) . DOI: 10.1061/(ASCE)0733-9437(2006)132:3(238)

Oliveira RZ, Oliveira LFC, Wehr TR, Borges LB, Bonomo R (2005) Comparative study of estimative models for reference evapotranspiration for the region of Goiânia, Go. Biosci. J. Uberlândia, 21(3):19-27.

Paes de Camargo, A. and Paes de Camargo, M.(2000) Numa revisão analítica da evapotranspiração potencial Bragantia. Campinas 59 2, pp. 125–137.

Penman, H.L. (1948) Natural evaporation from open water, bare soil, and grass. Proc. Roy. Soc. London A193:120-146.

Pereira DR, Yanagi SNM, Mello CR, Silva AM, Silva LA (2009) Performance of the reference evapotranspiration estimating methods for the Mantiqueira range region, MG, Brazil. Ciência Rural, Santa Maria 39(9):2488-2493.

Priestley CHB, Taylor RJ (1972) On the assessment of the surface heat flux and evaporation using large-scale parameters. Monthly Weather Review 100: 81–92

Popova Z, Kercheva M, Pereira LS (2006) Validation of the FAO methodology for computing ETo with limited data. Application to south Bulgaria. Irrig Drain. 55:201–215.

Rahimikhoob AR (2008) Comparative study of Hargreaves's and artificial neural network's methodologies in estimating reference evapotranspiration in a semiarid environment. Irrig Sci 26:253–259.

Reis EF, Bragança R, Garcia GO, Pezzopane JEM, Tagliaferre C (2007) Comparative study of the estimate of evaporate transpiration regarding the three locality state of Espirito Santo during the dry period. IDESIA (Chile) 25(3) 75-84

Samani Z (2000) Estimating solar radiation and evapotranspiration using minimum climatological data. J Irrig Drain Engin. 126(4):265–267.

Samani ZA, Pessarakli M (1986) Estimating potential crop evapotranspiration with minimum data in Arizona. Trans. ASAE (29): 522–524.

Sepaskhah, AR, Razzaghi, FH (2009) Evaluation of the adjusted Thornthwaite and Hargreaves-Samani methods for estimation of daily evapotranspiration in a semi-arid region of Iran. Archives of Agronomy and Soil Science, 55: 1, 51- 6

Sentelhas C, Gillespie TJ, Santos EA (2010) Evaluation of FAO Penman–Monteith and alternative methods for estimating reference evapotranspiration with missing data in Southern Ontario, Canada. Agricultural Water Management 97: 635–644.

Shahidian, S., Serralheiro, R.P., Teixeira, J.L., Santos, F.L., Oliveira, M.R.G., Costa, J.L., Toureiro, C.. Haie, N. (2007) Desenvolvimento dum sistema de rega automático, autónomo e adaptativo. I Congreso Ibérico de Agroingeneria.

Shahidian S. , Serralheiro R. , Teixeira J.L., Santos F.L., Oliveira M.R., Costa J., Toureiro C., Haie N., Machado R. (2009) Drip Irrigation using a PLC based Adaptive Irrigation System WSEAS Transactions on Environment and Development, Vol 2- Feb.

Shuttleworth, W.J., and I.R. Calder. (1979) Has the Priestley-Taylor equation any relevance to the forest evaporation? Journal of Applied Meteorology, 18: 639-646.

Smith, M, R.G. Allen, J.L. Monteith, L.S. Pereira, A. Perrier, and W.O. Pruitt. (1992) Report on the expert consultation on procedures for revision of FAO guidelines for prediction of crop water requirements. Land and Water Development Division, United Nations Food and Agriculture Service, Rome, Italy

Stanhill, G., 1961. A comparison of methods of calculating potential evapotranspiration from climatic data. Isr. J. Agric. Res.: Bet-Dagan (11), 159–171.

Teixeira JL, Shahidian S, Rolim J (2008) Regional analysis and calibration for the South of Portugal of a simple evapotranspiration model for use in an autonomous landscape irrigation controller.

Thornthwaite CW (1948) An approach toward a rational classification of climate. Geograph Rev.38:55–94.

Trajkovic S. (2005) Temperature-based approaches for estimating reference evapotranspiration. J Irrig Drain Engineer. 131(4):316–323

Trajkovic S (2007) Hargreaves versus Penman-Monteith under Humid Conditions Journal of Irrigation and Drainage Engineering, Vol. 133, No. 1, February 1.

Trajkovic, S, Stojnic, V. (2007) Effect of wind speed on accuracy of Turc Method in a humid climate. Facta Universitatis. 5(2):107-113

Xu, CY, Singh, VP (2000) Evaluation and Generalization of Radiation-based Methods for Calculating Evaporation, *Hydrolog. Processes* 14: 339–349.

Xu CY, Singh VP (2002) Cross Comparison of Empirical Equations for Calculating Potential evapotranspiration with data from Switzerland. Water Resources Management 16: 197-219.

5

Fuzzy-Probabilistic Calculations of Evapotranspiration

Boris Faybishenko
Lawrence Berkeley National Laboratory
Berkeley, CA
USA

1. Introduction

Evaluation of evapotranspiration uncertainty is needed for proper decision-making in the fields of water resources and climatic predictions (Buttafuoco et al., 2010; Or and Hanks, 1992; Zhu et al., 2007). However, in spite of the recent progress in soil-water and climatic uncertainty quantification, using stochastic simulations, the estimates of potential (reference) evapotranspiration (E_o) and actual evapotranspiration (ET) using different methods/models, with input parameters presented as PDFs or fuzzy numbers, is a somewhat overlooked aspect of water-balance uncertainty evaluation (Kingston et al., 2009). One of the reasons for using a combination of different methods/models and presenting the final results as fuzzy numbers is that the selection of the model is often based on vague, inconsistent, incomplete, or subjective information. Such information would be insufficient for constructing a single reliable model with probability distributions, which, in turn, would limit the application of conventional stochastic methods.

Several alternative approaches for modeling complex systems with uncertain models and parameters have been developed over the past ~50 years, based on fuzzy set theory and possibility theory (Zadeh, 1978; 1986; Dubois & Prade, 1994; Yager & Kelman, 1996). Some of these approaches include the blending of fuzzy-interval analysis with probabilistic methods (Ferson & Ginzburg, 1995; Ferson, 2002; Ferson et al., 2003). This type of analysis has recently been applied to hydrological research, risk assessment, and sustainable water-resource management under uncertainty (Chang, 2005), as well as to calculations of E_o, ET, and infiltration (Faybishenko, 2010).

The objectives of this chapter are to illustrate the application of a combination of probability and possibility conceptual-mathematical approaches — using fuzzy-probabilistic models — for predictions of potential evapotranspiration (E_o) and actual evapotranspiration (ET) and their uncertainties, and to compare the results of calculations with field evapotranspiration measurements.

As a case study, statistics based on monthly and annual climatic data from the Hanford site, Washington, USA, are used as input parameters into calculations of potential evapotranspiration, using the Bair-Robertson, Blaney-Criddle, Caprio, Hargreaves, Hamon, Jensen-Haise, Linacre, Makkink, Penman, Penman-Monteith, Priestly-Taylor, Thornthwaite, and Turc equations. These results are then used for calculations of evapotranspiration based on the modified Budyko (1974) model. Probabilistic calculations are performed using Monte

Carlo and p-box approaches, and fuzzy-probabilistic and fuzzy simulations are conducted using the RAMAS Risk Calc code. Note that this work is a further extension of this author's recently published work (Faybishenko, 2007, 2010).

The structure of this chapter is as follows: Section 2 includes a review of semi-empirical equations describing potential evapotranspiration, and a modified Budyko's model for evaluating evapotranspiration. Section 3 includes a discussion of two types of uncertainties — epistemic and aleatory uncertainties — involved in assessing evapotranspiration, and a general approach to fuzzy-probabilistic simulations by means of combining possibility and probability approaches. Section 4 presents a summary of input parameters and the results of E_0 and ET calculations for the Hanford site, and Section 5 provides conclusions.

2. Calculating potential evapotranspiration and evapotranspiration

2.1 Equations for calculations of potential evapotranspiration

The potential (reference) evapotranspiration E_0 is defined as evapotranspiration from a hypothetical 12 cm grass reference crop under well-watered conditions, with a fixed surface resistance of 70 s m^{-1} and an albedo of 0.23 (Allen et al., 1998). Note that this subsection includes a general description of equations used for calculations of potential evapotranspiration; it does not provide an analysis of the various advantages and disadvantages in applying these equations, which are given in other publications (for example, Allen et al., 1998; Allen & Pruitt, 1986; Batchelor, 1984; Maulé et al., 2006; Sumner & Jacobs, 2005; Walter et al., 2002).

The two forms of Baier-Robertson equations (Baier, 1971; Baier & Robertson, 1965) are given by:

$$E_0 = 0.157T_{max} + 0.158\,(T_{max} - T_{min}) + 0.109R_a - 5.39 \tag{1}$$

$$E_0 = -0.0039T_{max} + 0.1844(T_{max} - T_{min}) + 0.1136\,R_a + 2.811(e_s - e_a) - 4.0 \tag{2}$$

where E_0 = daily evapotranspiration (mm day^{-1}); T_{max} = the maximum daily air temperature, ^{o}C; T_{min} = minimum temperature, ^{o}C; R_a = extraterrestrial radiation (MJ m^{-2} day^{-1}) (ASCE 2005), e_s = saturation vapor pressure (kPa), and e_a = mean actual vapor pressure (kPa). Equation (1) takes into account the effect of temperature, and Equation (2) takes into account the effects of temperature and relative humidity.

The Blaney-Criddle equation (Allen & Pruitt, 1986) is used to calculate evapotranspiration for a reference crop, which is assumed to be actively growing green grass of 8–15 cm height:

$$E_0 = p\,(0.46\,\cdot T_{mean} + 8) \tag{3}$$

where E_0 is the reference (monthly averaged) evapotranspiration (mm day^{-1}), T_{mean} is the mean daily temperature ($^{\circ}C$) given as $T_{mean} = (T_{max} + T_{min})/2$, and p is the mean daily percentage of annual daytime hours.

The Caprio (1974) equation for calculating the potential evapotranspiration is given by

$$E_0 = 6.1\,\cdot 10^{-6}\,R_s\,[(1.8\,\cdot T_{mean}) + 1.0] \tag{4}$$

where E_0 = mean daily potential evapotranspiration (mm day^{-1}); R_s = daily global (total) solar radiation (kJ m^{-2} day^{-1}); and T_{mean} = mean daily air temperature ($^{\circ}C$).

The Hansen (1984) equation is given by:

$$E_o = 0.7 \Delta / (\Delta + \gamma) \cdot R_i / \lambda \tag{5}$$

where Δ = slope of the saturation vapor pressure vs. temperature curve, γ = psychrometric constant, R_i = global radiation, and λ = latent heat of water vaporization.
The Hargreaves equation (Hargreaves & Samani, 1985) is given by

$$E_o = 0.0023(T_{mean} + 17.8)(T_{max} - T_{min})^{0.5} R_a \tag{6}$$

where both E_o and R_a (extraterrestrial radiation) are in millimeters per day^{-1} (mm day^{-1}).
The Jensen and Haise (1963) equation is given by

$$E_o = R_s / 2450 [(0.025 T_{mean}) + 0.08] \tag{7}$$

where E_o = monthly mean of daily potential evapotranspiration (mm day^{-1}); R_s = monthly mean of daily global (total) solar radiation (kJ m^{-2} day^{-1}); and T_{mean} = monthly mean temperature.
The Linacre (1977) equation is given by:

$$E_o = [500 T_m / (100-L) + 15(T-Td)] / (80-T) \tag{8}$$

where E_o is in mm day^{-1}, T_m = temperature adjusted for elevation, $T_m = T + 0.006h$ (°C), h = elevation (m), T_d = dew point temperature (°C), and L = latitude (°).
The Makkink (1957) model is given by

$$E_o = 0.61 \Delta / (\Delta + \gamma) R_s / 2.45 - 0.12 \tag{9}$$

where R_s = solar radiation (MJ m^{-2} day^{-1}), and Δ and γ are the parameters defined above.
The Penman (1963) equation is given by

$$E_o = mR_n + \gamma 6.43(1+0.536 u_2) \delta e / \lambda_v (m + \gamma) \tag{10}$$

where Δ = slope of the saturation vapor pressure curve (kPa K^{-1}), R_n = net irradiance (MJ m^{-2} day^{-1}), ρ_a = density of air (kg m^{-3}), c_p = heat capacity of air (J kg^{-1} K^{-1}), δe = vapor pressure deficit (Pa), λ_v = latent heat of vaporization (J kg^{-1}), γ = psychrometric constant (Pa K^{-1}), and E_o is in units of kg/(m^{2}s).
The general form of the Penman-Monteith equation (Allen et al., 1998) is given by

$$E_o = [0.408 \Delta (R_n - G) + C_n \gamma / (T+273) u_2 (e_s-e_a)] / [\Delta + \gamma (1+C_d u_2)] \tag{11}$$

where E_o is the standardized reference crop evapotranspiration (in mm day^{-1}) for a short (0.12 m, with values C_n=900 and C_d=0.34) reference crop or a tall (0.5 m, with values C_n=1600 and C_d=0.38) reference crop, R_n = net radiation at the crop surface (MJ m^{-2} day^{-1}), G = soil heat flux density (MJ m^{-2} day^{-1}), T = air temperature at 2 m height (°C), u_2 = wind speed at 2 m height (m s^{-1}), e_s = saturation vapor pressure (kPa), e_a = actual vapor pressure (kPa), (e_s - e_a) = saturation vapor pressure deficit (kPa), Δ = slope of the vapor pressure curve (kPa °C^{-1}), and γ = psychrometric constant (kPa °C^{-1}).
The Priestley–Taylor (1972) equation is given by

$$E_o = \alpha 1/\lambda \Delta (R_n - G) / (\Delta + \gamma) \tag{12}$$

where λ = latent heat of vaporization (MJ kg^{-1}), R_n = net radiation (MJ m^{-2} day^{-1}), G = soil heat flux (MJ m^{-2} day^{-1}), Δ= slope of the saturation vapor pressure-temperature relationship (kPa °C^{-1}), γ = psychrometric constant (kPa °C^{-1}), and α = 1.26. Eichinger et al. (1996) showed that α=1.26 is practically constant for all typically observed atmospheric conditions and relatively insensitive to small changes in atmospheric parameters. (On the other hand, Sumner and Jacobs [2005] showed that α is a function of the green-leaf area index [LAI] and solar radiation.)

The Thornthwaite (1948) equation is given by

$$E_o = 1.6 \, (L/12) \, (N/30) \, (10 \, T_{mean \, (i)} / I)^\alpha \tag{13}$$

where E_o is the estimated potential evapotranspiration (cm/month), $T_{mean \, (i)}$ = average monthly (i) temperature (ºC); if $T_{mean \, (i)} < 0$, $Eo = 0$ of the month (i) being calculated, N = number of days in the month, L = average day length (hours) of the month being calculated, and I = heat index given by

$$I = \sum_{i=1}^{12} \left(\frac{T_{mean(i)}}{5} \right)^{1.514}$$

and $\alpha = (6.75 \cdot 10^{-7}) \, I^3 - (7.71 \cdot 10^{-5}) \, I^2 + (1.792 \cdot 10^{-2})I + 0.49239$

The Turc (1963) equation is given by

$$E_o = (0.0239 \cdot R_s + 50) \, [0.4/30 \cdot T_{mean} / (T_{mean} + 15.0)] \tag{14}$$

where E_o = mean daily potential evapotranspiration (mm/day); R_s = daily global (total) solar radiation (kJ/m^2/day); T_{mean} = mean daily air temperature (°C).

2.2 Modified Budyko's equation for evaluating evapotranspiration

For regional-scale, long-term water-balance calculations within arid and semi-arid areas, we can reasonably assume that (1) soil water storage does not change, (2) lateral water motion within the shallow subsurface is negligible, (3) the surface-water runoff and runon for regional-scale calculations simply cancel each other out, and (4) ET is determined as a function of the aridity index, ϕ: $ET=f(\phi)$, where $\phi = E_o/P$, which is the ratio of potential evapotranspiration, E_o, to precipitation, P (Arora 2002).

Budyko's (1974) empirical formula for the relationship between the ratio of ET/P and the aridity index was developed using the data from a number of catchments around the world, and is given by:

$$ET/P = \{\phi \tanh (1/\phi) \, [1 - \exp (-\phi)]\}^{0.5} \tag{15}$$

Equation (1) can also be given as a simple exponential expression (Faybishenko, 2010):

$$ET/P=a[1-\exp(-b\phi)] \tag{16}$$

with coefficients a =0.9946 and b =1.1493. The correlation coefficient between the calculations using (15) and (16) is R=0.999. Application of the modified Budyko's equation, given by an exponential function (2) with the ϕ value in single term, will simplify further calculations of ET.

3. Types of uncertainties in calculating evapotranspiration and simulation approaches

3.1 Epistemic and aleatory uncertainties

The uncertainties involved in predictions of evapotranspiration, as a component of soil-water balance, can generally be categorized into two groups—*aleatory* and *epistemic* uncertainties. Aleatory uncertainty arises because of the natural, inherent variability of soil and meteorological parameters, caused by the subsurface heterogeneity and variability of meteorological parameters. If sufficient information is available, probability density functions (PDFs) of input parameters can be used for stochastic simulations to assess aleatory evapotranspiration uncertainty. In the event of a lack of reliable experimental data, fuzzy numbers can be used for fuzzy or fuzzy-probabilistic calculations of the aleatory evapotranspiration uncertainty (Faybishenko 2010).

Epistemic uncertainty arises because of a lack of knowledge or poor understanding, ambiguous, conflicting, or insufficient experimental data needed to characterize coupled-physics phenomena and processes, as well as to select or derive appropriate conceptual-mathematical models and their parameters. This type of uncertainty is also referred to as subjective or reducible uncertainty, because it can be reduced as new information becomes available, and by using various models for uncertainty evaluation. Generally, variability, imprecise measurements, and errors are distinct features of uncertainty; however, they are very difficult, if not impossible, to distinguish (Ferson & Ginzburg, 1995).

In this chapter the author will consider the effect of aleatory uncertainty on evapotranspiration calculations by assigning the probability distributions of input meteorological parameters, and the effect of epistemic uncertainty is considered by using different evapotranspiration models.

3.2 Simulation approaches
3.2.1 Probability approach

A common approach for assessing uncertainty is based on Monte Carlo simulations, using PDFs describing model parameters. Another probability-based approach to the specification of uncertain parameters is based on the application of probability boxes (Ferson, 2002; Ferson et al., 2003). The probability box (p-box) approach is used to impose bounds on a cumulative distribution function (CDF), expressing different sources of uncertainty. This method provides an envelope of distribution functions that bounds all possible dependencies. An uncertain variable x expressed with a probability distribution, as shown in Figure 1a, can be represented as a variable that is bounded by a p-box $[\overline{F}, \underline{F}]$, with the right curve $\underline{F}(x)$ bounding the higher values of x and the lower probability of x, and the left curve $\overline{F}(x)$ bounding the lower values and the higher probability of x. With better or sufficiently abundant empirical information, the p-box bounds are usually narrower, and the results of predictions come close to a PDF from traditional probability theory.

3.2.2 Possibility approach

In the event of imprecise, vague, inconsistent, incomplete, or subjective information about models and input parameters, the uncertainty is captured using *fuzzy modeling theory*, or *possibility theory*, introduced by Zadeh (1978). For the past 50 years or so, possibility theory has successfully been applied to describe such systems as complex, large-scale engineering systems, social and economic systems, management systems, medical diagnostic processes, human perception, and others. The term *fuzziness* is, in general, used in possibility theory to

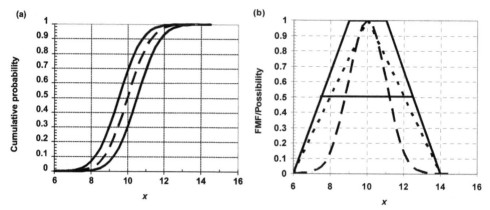

Fig. 1. Graphical illustration of uncertain numbers: (a) Cumulative normal distribution function (dashed line), with mean=10 and standard deviation σ=1, and a p-box — left bound with mean=9.5 and σ=0.9, and right bound with mean=10.5 and σ=1.1; and (b) Fuzzy trapezoidal (solid line) number, plotted using Eq. (17) with a=6, b=9, c=11, and d=14. Interval [b,c]=[9, 11] corresponds to FMF=1. Triangular (short dashes) and Gaussian (long dashes) fuzzy numbers are also shown. Figure (b) also shows an α-cut=0.5 (thick horizontal line) through the trapezoidal fuzzy number (Faybishenko 2010).

describe objects or processes that cannot be given precise definition or precisely measured. *Fuzziness* identifies a class (set) of objects with nonsharp (i.e., fuzzy) boundaries, which may result from imprecision in the meaning of a concept, model, or measurements used to characterize and model the system. Fuzzification implies replacing a set of crisp (i.e., precise) numbers with a set of fuzzy numbers, using fuzzy membership functions based on the results of measurements and perception-based information (Zadeh 1978). A fuzzy number is a quantity whose value is imprecise, rather than exact (as is the case of a single-valued number). Any fuzzy number can be thought of as a function whose domain is a specified set of real numbers. Each numerical value in the domain is assigned a specific "grade of membership," with 0 representing the smallest possible grade (full nonmembership), and 1 representing the largest possible grade (full membership). The grade of membership is also called the degree of possibility and is expressed using fuzzy membership functions (FMFs). In other words, a fuzzy number is a fuzzy subset of the domain of real numbers, which is an alternative approach to expressing uncertainty.

Several types of FMFs are commonly used to define fuzzy numbers: triangular, trapezoidal, Gaussian, sigmoid, bell-curve, Pi-, S-, and Z-shaped curves. As an illustration, Figure 1b shows a trapezoidal fuzzy number given by

$$f(x) = \begin{cases} 0, x \le a \\ \dfrac{x-a}{b-a}, a \le x \le b \\ 1, b \le x \le c \\ \dfrac{d-x}{d-c}, c \le x \le d \\ 0, d \le x \end{cases}, \tag{17}$$

where coefficients a, b, c, and d are used to define the shape of the trapezoidal FMF. When $a= b$, the trapezoidal number becomes a triangular fuzzy number.

Figure 1b also illustrates one of the most important attributes of fuzzy numbers, which is the notion of an α-cut. The α-cut interval is a crisp interval, limited by a pair of real numbers. An α-cut of 0 of the fuzzy variable represents the widest range of uncertainty of the variable, and an α-cut value of 1 represents the narrowest range of uncertainty of the variable.

Possibility theory is generally applicable for evaluating all kinds of uncertainty, regardless of its source or nature. It is based on the application of both hard data and the subjective (perception-based) interpretation of data. Fuzzy approaches provide a distribution characterizing the results of all possible magnitudes, rather than just specifying upper or lower bounds. Fuzzy methods can be combined with calculations of PDFs, interval numbers, or p-boxes, using the RAMAS Risk Calc code (Ferson 2002). In this paper, the RAMAS Risk Calc code is used to assess the following characteristic parameters of the fuzzy numbers and p-boxes:

- Mean — an interval between the means of the lower (left) and upper (right) bounds of the uncertain number x.
- Core — the most possible value(s) of the uncertain number x, i.e., value(s) with a possibility of one, or for which the probability can be any value between zero and one.
- Iqrange — an interval guaranteed to enclose the interquartile range (with endpoints at the 25th and 75th percentiles) of the underlying distribution.
- Breadth of uncertainty — for fuzzy numbers, given by the area under the membership function; for p-boxes, given by the area between the upper and lower bounds. The uncertainty decreases as the breadth of uncertainty decreases.

When fuzzy measures serve as upper bounds on probability measures, one could expect to obtain a conservative (bounding) prediction of system behavior. Therefore, fuzzy calculations may overestimate uncertainty. For example, the application of fuzzy methods is not optimal (i.e., it overestimates uncertainty) when sufficient data are available to construct reliable PDFs needed to perform a Monte Carlo analysis.

In a recent paper (Faybishenko 2010), this author demonstrated the application of the fuzzy-probabilistic method using a hybrid approach, with direct calculations, when some quantities can be represented by fuzzy numbers and other quantities by probability distributions and interval numbers (Kaufmann and Gupta 1985; Ferson 2002; Guyonnet et al. 2003; Cooper et al. 2006). In this paper, the author combines (aggregates) the results of Monte Carlo calculations with multiple E_0 models by means of fuzzy numbers and p-boxes, using the RAMAS Risk Calc software (Ferson 2002).

4. Hanford case study

4.1 Input parameters and modeling scenarios for the Hanford Site

The Hanford Site in Southeastern Washington State is one of the largest environmental cleanup sites in the USA, comprising 1,450 km^2 of semiarid desert. Located north of Richland, Washington, the Hanford Site is bordered on the east by the Columbia River and on the south by the Yakima River, which joins the Columbia River near Richland, in the Pasco Basin, one of the structural and topographic basins of the Columbia Plateau. The areal topography is gently rolling and covered with unconsolidated materials, which are sufficiently thick to mask the surface irregularities of the underlying material. Areas adjacent to the Hanford Site are primarily agricultural lands.

Meteorological parameters used to assign model input parameters were taken from the Hanford Meteorological Station (HMS—see http://hms.pnl.gov/), located at the center of the Hanford Site just outside the northeast corner of the 200 West Area, as well as from publications (DOE, 1996; Hoitink et al., 2002; Neitzel, 1996.) At the Hanford Site, the E_0 is estimated to be from 1,400 to 1,611 mm/yr (Ward et al. 2005), and the ET is estimated to be 160 mm/yr (Figure 2). A comparison of field estimates with the results of calculations performed in this paper is shown in Section 4.2. Calculations are performed using the temperature and precipitation time-series data representing a period of active soil-water balance (i.e., with no freezing) from March through October for the years 1990–2007. A set of meteorological parameters is summarized in Table 1, which are then used to develop the input PDFs and fuzzy numbers shown in Figure 3.

Several modeling scenarios were developed (Table 2) to assess how the application of different models for input parameters affects the uncertainty of E_0 and ET calculations. For the sake of simulation simplicity, the input parameters are assumed to be independent variables. Scenarios 0 to 8, described in detail in Faybishenko (2010), are based on the application of a single Penman model for E_0 calculations, with annual average values of input parameters. Scenario 0 was modeled using input PDFs by means of Monte Carlo simulations, using RiskAMP Monte Carlo Add-In Library version 2.10 for Excel. Scenarios 1 through 8 were simulated by means of the RAMAS Risk Calc code. Scenario 1 was simulated using input PDFs, and the results are given as p-box numbers. Scenarios 2 through 6 were simulated applying both PDFs and fuzzy number inputs, corresponding to α-cuts from 0 to 1). Scenarios 7 and 8 were simulated using only fuzzy numbers. The calculation results of Scenarios 0 through 8 are compared in this chapter with newly calculated Scenarios 9 and 10, which are based on Monte Carlo calculations by means of all E_0 models, described in Section 2, and then bounding the resulting PDFs by a trapezoidal fuzzy number (Scenario 9) and the p-box (Scenario 10).

Type of data	Parameters		Wind speed (km/hr)	Relative humidity (%)		Albedo	Solar radiation (Ly/day)	Annual precipi- tation (mm/yr)	Temperature (°C)	
				Max	Min				Max	Min
PDFs	Mean		15.07	80.2	33.3	0.21	332.55	185	33.41	2.87
	Standard Deviation		0.92	4.01	1.66	0.021	16.63	55.62	1.08	1.11
Trape- zoidal FMFs	α= 0	Min	12.31	68.17	28.29	0.15	282.66	46.0	30.17	0.0
		Max	17.84	92.23	38.31	0.27	382.44	324.1	36.65	6.17
	α=1	Min	14.61	78.2	32.47	0.22	324.24	157.2	32.87	2.32
		Max	15.53	82.2	34.14	0.27	382.44	212.8	33.95	3.42

Table 1. Meteorological parameters from the Hanford Meteorological Station used for E_0 calculations for all scenarios (the data sources are given in the text).

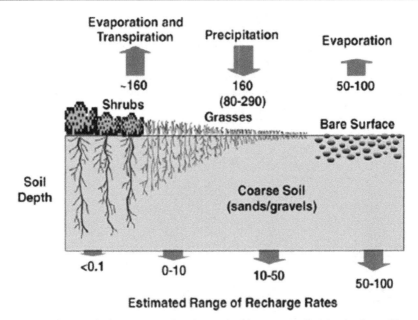

Fig. 2. Estimated water balance *ET* and recharge/infiltration at the Hanford site (Gee et al, 2007).

Scenarios	Input parameters						Output
	Wind speed	Humidity	Albedo	Solar radiation	Precipitation	Temperature	para-meters
0	PDF	PDF	PDF	PDF	PDF	PDF	PDF
1	PDF	PDF	PDF	PDF	PDF	PDF	p-box
2	Fuzzy	PDF	PDF	PDF	PDF	PDF	Hybrid
3	Fuzzy	Fuzzy	PDF	PDF	PDF	PDF	Hybrid
4	Fuzzy	Fuzzy	Fuzzy	PDF	PDF	PDF	Hybrid
5	Fuzzy	Fuzzy	Fuzzy	Fuzzy	PDF	PDF	Hybrid
6	Fuzzy	Fuzzy	Fuzzy	Fuzzy	Fuzzy	PDF	Hybrid
7[1]	Fuzzy	Fuzzy	Fuzzy	Fuzzy	Fuzzy	Fuzzy	Fuzzy
8[2]	Fuzzy	Fuzzy	Fuzzy	Fuzzy	Fuzzy	Fuzzy	Fuzzy
9[3]	PDF	PDF	PDF	PDF	PDF	PDF	Fuzzy
10[3]	PDF	PDF	PDF	PDF	PDF	PDF	p-box

Notes:
[1] In Scenario 7, all FMFs are trapezoidal.
[2] In Scenario 8, all FMFs are triangular: the mean values of parameters, which are given in Table 1, are used for $\alpha=1$; and the minimum and maximum values of parameters, given in Table 1 for trapezoidal FMFs (Scenario 7), are also used for $\alpha=0$ of triangular FMFs in Scenario 8.
[3] In Scenarios 9 and 10, input parameters are monthly averaged.

Table 2. Scenarios of input and output parameters used for water-balance calculations (Scenarios 0, and 1-8 are from Faybishenko, 2010).

4.2 Results and comparison with field data
4.2.1 Potential evapotranspiration (E_o)
Figure 4a shows cumulative distributions of E_o from different models, along with an aggregated p-box, and Figure 4b shows the corresponding FMFs (calculated as normalized PDFs) of E_o from different models, along with an aggregated trapezoidal fuzzy E_o. These figures illustrate that the Baier-Robertson (Eq. 1), Blaney-Criddle (Eq. 3), Hargreaves (Eq. 6), Penman (Eq. 10), Penman-Monteith (Eq. 11) (for tall plants), and Priestly-Taylor (Eq. 12) models provide the best match with field data, while the Makkink (Eq. 9) and Thornthwaite (Eq. 13) models significantly underestimate the E_o, and the Linacre (Eq. 8) and Baier-Robertson (Eq. 2) models greatly overestimate E_o.

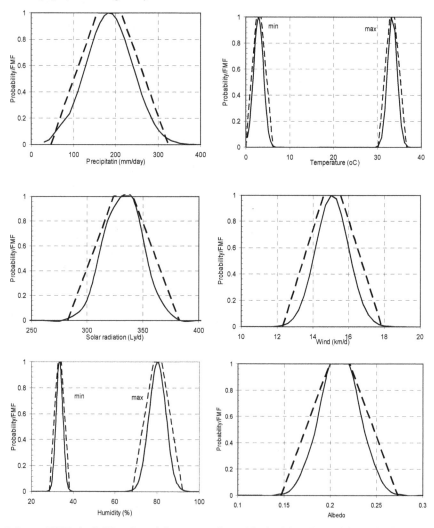

Fig. 3. Input PDFs (solid lines) and fuzzy numbers (dashed lines) used for calculations (Faybishenko, 2010).

Figure 5a demonstrates that the E_0 mean from Monte Carlo simulations is within the mean ranges from the p-box (Scenario 1) and fuzzy-probabilistic scenarios (Scenarios 2-6). It also corresponds to a midcore of the fuzzy scenario with trapezoidal FMFs (Scenario 7), the core of the fuzzy scenario with triangular FMFs (Scenario 8), and the centroid values of the fuzzy E_0 of Scenario 9, as well as a p-box of Scenario 10.

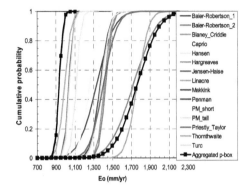
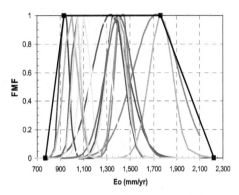

Fig. 4. (a) Cumulative probability of potential evapotranspiration calculated using different E_0 formulae; an aggregated p-box, which is shown by a black line with solid squares: normal distribution with the left/minimum curve—mean=933, var=1070, and the right/max curve—mean=1763, var=35755; and (b) corresponding fuzzy numbers (calculated from normalized PDFs); an aggregated trapezoidal fuzzy number is shown by a black line—Eq. (17) with a=772, b=933, c=1763, and d=2222. (all numbers of E_0 are in mm/yr)

The range of means from the p-box and fuzzy-probabilistic calculations for α=1 is practically the same, indicating that including fuzziness within the input parameters does not change the range of most possible E_0 values. Figure 5a shows that the core uncertainty of the trapezoidal FMFs (Scenario 7) is the same as the uncertainty of means for fuzzy-probabilistic calculations for α =1. Obviously, the output uncertainty decreases for the input triangular FMFs (Scenario 8), because these FMFs resemble more tightly the PDFs used in other scenarios. Figure 5a also illustrates that a relatively narrow range of field estimates of E_0— from 1,400 to 1,611 mm/yr for the Hanford site (Ward 2005)—is well within the calculated uncertainty of E_0 values. Note from Figure 5a that the uncertainty ranges from p-box, hybrid, and fuzzy calculations significantly exceed those from Monte Carlo simulations for a single Penman model, but are practically the same as those from calculations using multiple E_0 models.

Characteristic parameters (Figures 5a) and the breadth of uncertainty (Figure 6a) of E_0 calculated from multiple models—Scenarios 9 and 10—are in a good agreement with field measurements and other calculation scenarios.

4.2.2 Evapotranspiration (ET)

Figure 5b shows that the mean ET of ~184 mm/yr from Monte Carlo simulations (Scenario 0) is practically the same as the ET means for Scenarios 1 through 5 and the core value for Scenario 8. The greater ET uncertainty for Scenario 6 (precipitation is simulated using a fuzzy number) can be explained by the relatively large precipitation range for α=0—from 46 to 324 mm/yr. At the same time, the means of ET values for α =1 range within relatively narrow limits, as the precipitation for α =1 changes from 157.2 to 212.8 mm/yr (see Table 1).

The breadth of uncertainty of *ET* (Figure 6b) is practically the same for Scenarios 1 through 5, increase for Scenarios 6, 7, and 8 in the account of calculations using a fuzzy precipitation, and then decrease for Scenarios 9 and 10 using multiple E_o models. A smaller range of *ET* uncertainty calculated using multiple E_o models can be explained by the fact that the Budyko curve asymptotically reaches the limit of *ET*/*P*=1 for high values of the aridity index, which are typical for the semi-arid climatic conditions of the Hanford site.

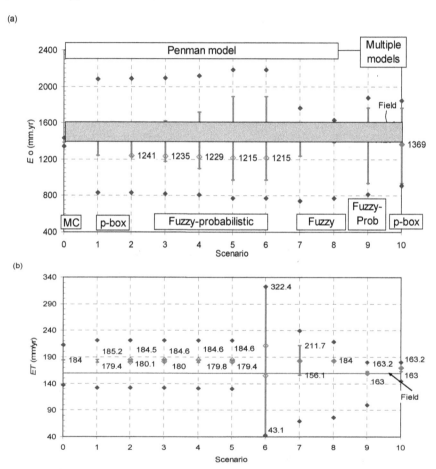

Fig. 5. Results of calculations of E_o (a) and *ET* (b) and comparison with field measurements. Red vertical lines are the mean intervals (Scenarios 1-6, and 10) and core intervals (Scenarios 7, 8, and 9), the blue diamonds indicate the interquartile ranges with endpoints at the 25th and 75th percentiles of the underlying distribution. Red open diamonds for Scenarios 2-6 indicate the mean intervals for the hybrid level=10 (Faybishenko 2010), and red solid diamonds for Scenarios 7-10 indicate centroid values. The height of a shaded area in figure *a* indicates the range of E_o from field measurements. (Results of calculations of Scenarios 0-8 are from Faybishenko, 2010.)

The calculated means for Scenarios 0, 1-5, and 8 exceed the field estimates of ET of 160 mm/yr (Gee et al., 1992; 2007) by 22 to 24 mm/yr. This difference can be explained by Gee et al. using a lower value of annual precipitation (160 mm/yr for the period prior to 1990) in their calculations, while our calculations are based on using a greater mean annual precipitation (185 mm/yr), averaged for the years from 1990 to 2007. The field-based data are within the ET uncertainty range for Scenarios 6 and 7, since the precipitation range is wider for these scenarios. Calculations using multiple E_o models generated the ET values (Scenarios 9 and 10), which are practically the same as those from field measurements.

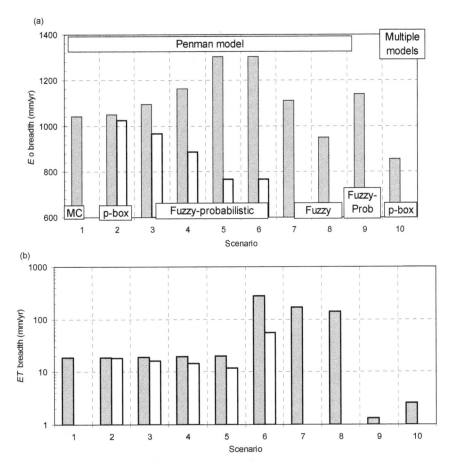

Fig. 6. Breadth of uncertainty of E_o and ET. For Scenarios 2-6, grey and white bars indicate the maximum and minimum uncertainty, correspondingly. (Results of calculations of Scenarios 0-8 are from Faybishenko, 2010.)

5. Conclusions

The objectives of this chapter are to illustrate the application of a fuzzy-probabilistic approach for predictions of E_o and ET, and to compare the results of calculations with those from field measurements at the Hanford site. Using historical monthly averaged data from the Hanford Meteorological Station, this author employed Monte-Carlo simulations to assess the frequency distribution and statistics of input parameters for these models, which are then used as input into probabilistic simulations. The effect of aleatory uncertainty on calculations of evapotranspiration is assessed by assigning the probability distributions of input meteorological parameters, and the combined effect of aleatory and epistemic (model) uncertainty is then expressed by means of aggregating the results of calculations using a p-box and fuzzy numbers. To illustrate the application of these approaches, the potential evapotranspiration is calculated using the Bair-Robertson, Blaney-Criddle, Caprio, Hargreaves-Samani, Hamon, Jensen-Haise, Linacre, Makkink, Priestly-Taylor, Penman, Penman-Monteith, Thornthwaite, and Turc models, and evapotranspiration is then determined based on the modified Budyko (1974) model. Probabilistic and fuzzy-probabilistic calculations using multiple E_0 models generate the E_0 and ET results, which are well within the range of field measurements and the application of a single Penman model. The Baier-Robertson, Blaney-Criddle, Hargreaves, Penman, Penman-Monteith, and Priestly-Taylor models provide the best match with field data.

6. Acknowledgment

This work was partially supported by the Director, Office of Science, Office of Biological and Environmental Remediation Sciences of the U.S. Department of Energy, and the DOE EM-32 Office of Soil and Groundwater Remediation (ASCEM project) under Contract No. DE-AC02-05CH11231 to Lawrence Berkeley National Laboratory.

7. References

Allen, R.G. & Pruitt, W.O. (1986), Rational Use of the FAO Blaney-Criddle Formula, *Journal of Irrigation and Drainage Engineering*, Vol. 112, No. 2, pp. 139-155, doi 10.1061/(ASCE)0733-9437(1986)112:2(139)

Allen, R.G.; Pereira L.S.; Raes, D. & Smith, M. (1998). Crop evapotranspiration - Guidelines for computing crop water requirements - FAO Irrigation and drainage paper 56.

Arora V.K. (2002). The use of the aridity index to assess climate change effect on annual runoff, *J. of Hydrology*, Vol. 265, pp. 164–177.

ASCE (2005). The ASCE Standardized Reference Evapotranspiration Equation. Edited by R. G. Allen, I.A. Walter, R.; Elliott, T. Howell, D. Itenfisu, and M. Jensen. New York, NY: American Society of Civil Engineers.

Baier, W. & Robertson, G.W. 1965. Estimation of latent evaporation from simple weather observations. Can. J. Soil Sci. 45, pp. 276-284.

Baier, W. (1971). Evaluation of latent evaporation estimates and their conversion to potential evaporation. *Can. J. of Plant Sciences* 51, pp. 255-266.

Budyko, M.I. (1974).*Climate and Life*, Academic, San Diego, Calif., 508 pp.

Buttafuoco, G.; Caloiero, T.; & Coscarelli, R. (2010). Spatial uncertainty assessment in modelling reference evapotranspiration at regional scale, *Hydrol. Earth Syst. Sci.*, 14, pp. 2319-2327, doi:10.5194/hess-14-2319-2010, 2010.

Caprio, J.M. (1974). The Solar Thermal Unit Concept in Problems Related to Plant Development and Potential Evapotranspiration. In: H. Lieth (Editor), *Phenology and Seasonality Modeling. Ecological Studies.* Springer Verlag, New York, pp. 353-364.

Chang N-B (2005). Sustainable water resources management under uncertainty, *Stochast Environ Res and Risk Assess*, 19, pp. 97–98.

Cooper J.A.; Ferson, S.; & Ginzburg, L. (2006). Hybrid processing of stochastic and subjective uncertainty data, *Risk Analysis*, Vol. 16, No. 6, pp. 785–791.

DOE (1996). Final Environmental Impact Statement for the Tank Waste Remediation System, Hanford Site, Richland, Washington, DOE/EIS-0189. Available from http://www.globalsecurity.org/wmd/library/report/enviro/eis-0189/app_i_3.htm

Dubois, D. & Prade, H. (1994). Possibility theory and data fusion in poorly informed environments. *Control Engineering Practice* 2(5), pp. 811-823.

Eichinger W.E.; Parlange, M.B. & Stricker H. (1996). On the Concept of Equilibrium Evaporation and the Value of the Priestley-Taylor Coefficient, *Water Resour. Res.*, 32(1): 161-164, doi:10.1029/95WR02920.

Faybishenko, B. (2007). Climatic Forecasting of Net Infiltration at Yucca Mountain Using Analogue Meteorological Data, *Vadose Zone Journal*, 6: 77–92.

Faybishenko, B. (2010), Fuzzy-probabilistic calculations of water-balance uncertainty, *Stochastic Environmental Research and Risk Assessment*, Vol. 24, No. 6, pp. 939–952.

Ferson, S. (2002). RAMAS Risk Calc 4.0 Software: Risk assessment with uncertain numbers, CRC Press.

Ferson, S. & Ginzburg, L. (1995) Hybrid arithmetic. *Proceedings of the 1995 Joint ISUMA/NAFIPS Conference*, IEEE Computer Society Press, Los Alamitos, California, pp. 619-623.

Ferson, S.; Kreinovich, V.; Ginzburg, L; Myers, D.S. & Sentz, K. (2003). Constructing probability boxes and Dempster-Shafer structures, SAND REPORT, SAND2002-4015.

Gee, G.W.; Fayer, M.J.; Rockhold, M.L. & Campbell, M.D. (1992). Variations in recharge at the Hanford Site. *Northwest Sci.* 66, pp. 237–250.

Gee, G.W.; Oostrom, M.; Freshley, M.D.; Rockhold, M.L. & Zachara, J.M. (2007). Hanford site vadose zone studies: An overview, *Vadose Zone Journal* Vol. 6, pp. 899-905.

Guyonne, D.; Dubois, D. ; Bourgine, B.; Fargier, H.; Côme, B. & Chilès, J.-P. (2003). Hybrid method for addressing uncertainty in risk assessments. *Journal of Environmental Engineering* 129: 68-78.

Hansen, S. (1984). Estimation of Potential and Actual Evapotranspiration, *Nordic Hydrology,* 15, 1984, pp. 205-212, Paper presented at the Nordic Hydrological Conference (Nyborg, Denmark, August 1984).

Hargreaves, G.H. & Z.A. Samani (1985). Reference crop evapotranspiration from temperature. *Transaction of ASAE* 1(2), pp. 96-99.

Hoitink, D.J.; Burk, K.W ; Ramsdell, Jr, J.V; & Shaw, W.J. (2003). Hanford Site Climatological Data Summary 2002 with Historical Data . PNNL-14242, Pacific Northwest National Laboratory, Richland, WA.

Jensen, M.E. & Haise, H.R. (1963). Estimating evapotranspiration from solar radiation. *J. Irrig. Drainage Div. ASCE*, 89: 15-41.

Kaufmann, A. & Gupta, M.M. (1985). *Introduction to Fuzzy Arithmetic*, New York: Van Nostrand Reinhold.

Kingston, D.G.; Todd, M.C.; Taylor, R.G.; Thompson, J.R. & Arnell N.W. (2009). Uncertainty in the estimation of potential evapotranspiration under climate change, *Geophysical Research Letters*, Vol. 36, L20403. doi:10.1029/2009GL040267

Linacre, E. T. (1977). A simple formula for estimating evaporation rates in various climates, using temperature data alone. *Agric. Meteorol.*, 18, pp. 409--424.

Makkink, G. F. (1957). Testing the Penman formula by means of lysimiters, *J. Institute of Water Engineering*, 11, pp. 277-288.

Maulé C.; Helgason, W.; McGin, S. & Cutforth, H. (2006). Estimation of standardized reference evapotranspiration on the Canadian Prairies using simple models with limited weather data. *Canadian Biosystems Engineering* 48, pp. 1.1 - 1.11.

Neitzel, D.A. (1996) Hanford Site National Environmental Policy Act (NEPA) Characterization. PNL-6415, Rev. 8. Pacific Northwest National Laboratory. Richland, Washington.

Or D. & Hanks, R.J. (1992). Spatial and temporal soil water estimation considering soil variability and evapotranspiration uncertainty, *Water Resour. Res.* Vol. 28, No. 3, pp. 803-814. doi:10.1029/91WR02585

Penman H.L (1963). *Vegetation and hydrology*. Tech. Comm. No. 53, Commonwealth Bureau of Soils, Harpenden, England. 125 pp.

Priestley, C.H.B. & Taylor, R.J. (1972). On the assessment of surface heat flux and evaporation using large-scale parameters. *Mon. Weather Rev.* 100(2), pp. 81–92.

Sumner D.M. & Jacobs, J.M. (2005). Utility of Penman–Monteith, Priestley–Taylor, reference evapotranspiration, and pan evaporation methods to estimate pasture evapotranspiration, *Journal of Hydrology*, 308, pp. 81–104.

Thornthwaite, C.W. (1948). An approach toward a rational classification of climate. *Geogr. Rev.* 38, pp. 55–94.

Turc, L. (1963). Evaluation des besoins en eau d'irrigation, évapotranspiration potentielle, formulation simplifié et mise à jour. *Ann. Agron.*, 12: 13-49.

Walter, I.A.; Allen, R.G.; Elliott, R.; Itenfisu, D.; Brown, P.; Jensen, M.E.; Mecham, B.; Howell, T.A.; Snyder, R.L.; Eching, S.; Spofford, T.; Hattendorf, M.; Martin, D.; Cuenca, R.H. & Wright, J.L. (2002). The ASCE standardized reference evapotranspiration equation. *Rep. Task Com. on Standardized Reference Evapotranspiration July 9, 2002, EWRI-Am. Soc. Civil. Engr.*, Reston, VA, 57 pp. /w six Appendices. http://www.kimberly.uidaho.edu/water/asceewri/main.pdf.

Ward, A.L.; Freeman, E.J.; White, M.D. & Zhang, Z.F. (2005). STOMP: Subsurface Transport Over Multiple Phases, Version 1.0, Addendum: Sparse Vegetation Evapotranspiration Model for the Water-Air-Energy Operational Mode, PNNL-15465.

Yager. R. & Kelman, A. (1996). Fusion of fuzzy information with considerations for compatibility, partial aggregation, and reinforcement. *International Journal of Approximate Reasoning*, 15(2), pp. 93-122.

Zadeh, L. (1978). Fuzzy sets as a basis for a theory of possibility. *Fuzzy Sets and Systems*, 1, pp. 3-28.

Zadeh, L.A. (1986). A Simple view of the Dempster-Shafer theory of evidence and its implication for the rule of combination. *The AI Magazine* 7, pp. 85-90.

Zhu J.; Young, M.H. & Cablk, M.E. (2007) Uncertainty Analysis of Estimates of Ground-Water Discharge by Evapotranspiration for the BARCAS Study Area, DHS Publication No. 41234.

6

Impact of Irrigation on Hydrologic Change in Highly Cultivated Basin

Tadanobu Nakayama[1,2]
¹National Institute for Environmental Studies (NIES)
16-2 Onogawa, Tsukuba, Ibaraki
²Centre for Ecology & Hydrology (CEH)
Crowmarsh Gifford, Wallingford, Oxfordshire
¹Japan
²United Kingdom

1. Introduction

With the development of regional economies, the water use environment in the Yellow River Basin, China, has changed greatly (Fig. 1). The river is well known for its high sediment content, frequent floods, unique channel characteristics in the downstream (where the river bed lies above the surrounding land), and limited water resources. This region is heavily irrigated, and combinations of increased food demand and declining water availability are creating substantial pressures. Some research emphasized human activities such as irrigation water withdrawals dominate annual streamflow changes in the downstream in addition to climate change (Tang et al., 2008a). The North China Plain (NCP), located in the downstream area of the Yellow River, is one of the most important grain cropping areas in China, where water resources are also the key to agricultural development, and the demand for groundwater has been increasing. Groundwater has declined dramatically over the previous half century due to over-pumping and drought, and the area of saline-alkaline land has expanded (Brown and Halweil, 1998; Shimada, 2000; Chen et al., 2003b; Nakayama et al., 2006).

Since the completion of a large-scale irrigation project in 1969, noticeable cessation of flow has been observed in the Yellow River (Yang et al., 1998; Fu et al., 2004) resulting from intense competition between water supply and demand, which has occurred increasingly often. The ratio of irrigation water use (defined as the ratio of the annual gross use for irrigation relative to the annual natural runoff) having increased continuously from 21% to 68% during the last 50 years, indicating that the current water shortage is closely related to irrigation development (Yang et al., 2004a). This shortage also reduces the water renewal time (Liu et al., 2003) and renewability of water resources (Xia et al., 2004). This has been accompanied by a decrease in precipitation in most parts of the basin (Tang et al., 2008b). To ensure sustainable water resource use, it is also important to understand the contributions of human intervention to climate change in this basin (Xu et al., 2002), in addition to clarifying the rather complex and diverse water system in the highly cultivated region.

The objective of this research is to clarify the impact of irrigation on the hydrologic change in the Yellow River Basin, an arid to semi-arid environment with intensive cultivation.

Combination of the National Integrated Catchment-based Eco-hydrology (NICE) model (Nakayama, 2008a, 2008b, 2009, 2010, 2011a, 2011b; Nakayama and Fujita, 2010; Nakayama and Hashimoto, 2011; Nakayama and Watanabe, 2004, 2006, 2008a, 2008b, 2008c; Nakayama et al., 2006, 2007, 2010, 2011) with complex components such as irrigation, urban water use, and dam/canal systems has led to the improvement in the model, which simulates the balance of both water budget and energy in the entire basin with a resolution of 10 km. The simulated results also evaluates the complex hydrological processes of river dry-up, agricultural/urban water use, groundwater pumping, and dam/canal effects, and to reveal the impact of irrigation on both surface water and groundwater in the basin. This approach will help to clarify how the substantial pressures of combinations of increased food demand and declining water availability can be overcome, and how effective decisions can be made regarding sustainable development under sound socio-economic conditions in the basin.

2. Material and methods

2.1 Coupling of process-based model with complex irrigation procedures
Previously, the author developed the process-based NICE model, which includes surface-unsaturated-saturated water processes and assimilates land-surface processes describing the variations of LAI (leaf area index) and FPAR (fraction of photosynthetically active radiation) from satellite data (Fig. 2) (Nakayama, 2008a, 2008b, 2009, 2010, 2011a, 2011b; Nakayama and Fujita, 2010; Nakayama and Hashimoto, 2011; Nakayama and Watanabe, 2004, 2006, 2008a, 2008b, 2008c; Nakayama et al., 2006, 2007, 2010, 2011). The unsaturated layer divides canopy into two layers, and soil into three layers in the vertical dimension in the SiB2 (Simple Biosphere model 2) (Sellers et al., 1996). About the saturated layer, the NICE solves three-dimensional groundwater flow for both unconfined and confined aquifers. The hillslope hydrology can be expressed by the two-layer surface runoff model including freezing/thawing processes. The NICE connects each sub-model by considering water/heat fluxes: gradient of hydraulic potentials between the deepest unsaturated layer and the groundwater, effective precipitation, and seepage between river and groundwater.

In an agricultural field, NICE is coupled with DSSAT (Decision Support Systems for Agro-technology Transfer) (Ritchie et al., 1998), in which automatic irrigation mode supplies crop water requirement, assuming that average available water in the top layer falls below soil moisture at field capacity for cultivated fields (Nakayama et al., 2006). The model includes different functions of representative crops (wheat, maize, soybean, and rice) and simulates automatically dynamic growth processes. Potential evaporation is calculated on Priestley and Taylor equation (Priestley and Taylor, 1972), and plant growth is based on biomass formulation, which is limited by various reduction factors like light, temperature, water, and nutrient, et al. (Nakayama et al., 2006; Nakayama and Watanabe, 2008b; Nakayama, 2011a).

In this study, the NICE was coupled with complex sub-systems in irrigation and dam/canal in order to develop coupled human and natural systems and to analyze impact of irrigation on hydrologic change in highly cultivated basin. The return flow was evaluated from surface drainage and from groundwater, whereas previous studies had considered only surface drainage (Liu et al., 2003; Xia et al., 2004; Yang et al., 2004a). The gross loss of river water to irrigation includes losses via canals and leakage into groundwater in the field, and can be estimated as the difference between intake from, and return to the river.

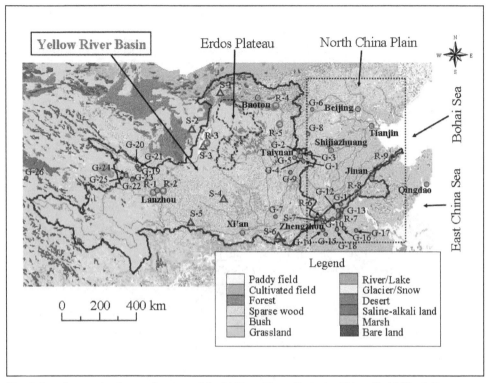

Fig. 1. Land cover in the study area of the Yellow River Basin in China. Bold black line shows the boundary of the basin. Black dotted line is the border of the North China Plain (NCP), which includes the downstream of the Yellow River Basin. Verification data are also plotted in this figure: river discharge (open red circle), soil moisture (open brown triangle), and groundwater level (green dot).

Irrigation withdrawals in the basin account for about 90% of total surface abstraction and 60% of groundwater withdrawal (Chen et al., 2003a). The model was improved for application to irrigated fields where water is withdrawn from both groundwater and river, and therefore the river dry-up process can be reproduced well. As the initial conditions, the ratios of river to aquifer irrigation were set at constant values. In the calibration procedure, these values were changed from initial conditions in order to reproduce the observation data as closely as possible after repeated trial and error (Oreskes et al., 1994). A validation procedure was then conducted in order to confirm the simulation under the same set of parameters, which resulted into reproducing reasonably the observed values. Spring/winter wheat, summer maize, and summer rice were automatically simulated in sequence analysis mode in succession by inputting previous point data for each crop type (Wang et al., 2001; Liu et al., 2002; Tao et al., 2006) and spatial distribution data (Chinese Academy of Sciences, 1988; Fang et al., 2006). The deficit water in the irrigated fields was automatically withdrawn and supplied from the river or the aquifer in the model in order to satisfy the observed hydrologic variables like soil moisture, river discharge, groundwater level, LAI, evapotranspiration, and crop coefficient. So, NICE simulates drought impact and includes the effect of water stress implicitly. Details are given in the previous researches (Nakayama et al., 2006; Nakayama and Watanabe, 2008b; Nakayama, 2011a, 2011b).

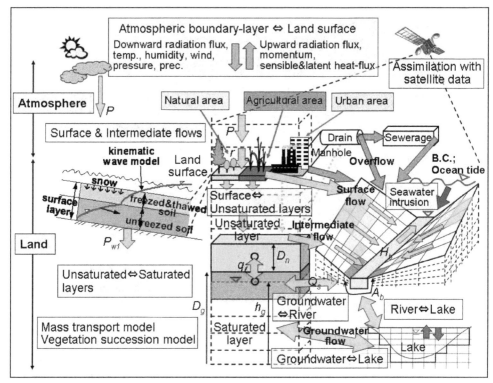

Fig. 2. National Integrated Catchment-based Eco-hydrology (NICE) model.

Another important characteristics of the study area is that there are many dams and canals to meet the huge demand for agricultural, industrial, and domestic water use (Ren et al., 2002) (Fig. 1), and exist six large dams on the main river (Yang et al., 2004a). Because there are few available data on discharge control at most of these dams, the model uses a constant ratio of dam inflow to outflow, which is a simpler approach than that of the storage-runoff function model (Sato et al., 2008). There are also many complex canals in the three large irrigation zones (Qingtongxia in Ningxia Hui, Hetao in Inner Mongolia, and Weisan in Shandong Province), in addition to the NCP, making it very difficult to evaluate the flow dynamics there. Because it is impossible to obtain the observed discharge and data related to the control of the weir/gate at every canal, it is effective to estimate the flow dynamics only in main canals as the first approximation when attempting to evaluate the hydrologic cycle in the entire basin in the same way as (Nakayama et al., 2006; Nakayama, 2011a). Therefore, NICE simulates the discharge only in a main canal assuming that this is defined as the difference in hydraulic potentials at both junctions similar to the stream junction model (Nakayama and Watanabe, 2008b). The dynamic wave effect is also important for the simulation of meandering rivers and smaller slopes, because the backwater effect is predominant (Nakayama and Watanabe, 2004). When a river flow is very low and almost zero at some point in the simulation, the dynamic wave theory requires a lot more computation time and sometimes becomes unstable. Therefore, the model applies a threshold water level of 1 mm to ensure simulation stability and to include the dry-up process. The model also includes the seepage process which is decided by some parameters such as hydraulic conductivity of the river bed, cross-sectional area of the groundwater section, and river bed thickness. Details are described in Nakayama (2011b).

2.2 Model input data and running the simulation

Six-hour reanalysed data for downward radiation, precipitation, atmospheric pressure, air temperature, air humidity, wind speed at a reference level, FPAR, and LAI were input into the model after interpolation of ISLSCP (International Satellite Land Surface Climatology Project) data with a resolution of 1° x 1° (Sellers et al., 1996) in inverse proportion to the distance back-calculated in each grid. Because the ISLSCP precipitation data had the least reliability and underestimated the observed values at peak times, rain gauge daily precipitation data collected at 3,352 meteorological stations throughout the study area were used to correct the ISLSCP precipitation data. Mean elevation of each 10-km grid cell was calculated from the spatial average of a global digital elevation model (DEM; GTOPO30) with a horizontal grid spacing of 30 arc–seconds (~1 km) (USGS, 1996). Digital land cover data produced by the Chinese Academy of Sciences (CAS) based on Landsat TM data from the early 1990s (Liu, 1996) were categorized for the simulation (Fig. 1). Vegetation class and soil texture were categorized and digitized into 1-km mesh data by using 1:4,000,000 and 1:1,000,000 vegetation and soil maps of China (Chinese Academy of Sciences, 1988, 2003). The author's previous research showed that these finer-resolution products are highly correlated with the ISLSCP (Nakayama, 2011b). The geological structure was divided into four types on the basis of hydraulic conductivity, the specific storage of porous material, and specific yield by scanning and digitizing the geological material (Geological Atlas of China, 2002) and core–sampling data at some points (Zhu, 1992).

The irrigation area was calculated from the GIS data based on Landsat TM data from the early 1990s (Liu, 1996), and the calculated value agrees well with the previous results from that period (Yang et al., 2004a) (Table 1), as described in Nakayama (2011b). Most of the irrigated fields are distributed in the middle and lower regions of the Yellow River mainstream and in the NCP (Fig. 3). The agricultural areas in the upper regions and Erdos Plateau are dominated by dryland fields. Spring/winter wheat was predominant in the upper and middle of the arid and semi-arid regions, and double cropping of winter wheat and summer maize was usually practised in the middle and downstream and in the NCP's relatively warm and humid environment (Wang et al., 2001; Liu et al., 2002; Fang et al., 2006; Nakayama et al., 2006; Tao et al., 2006). The averaged water use during 1987-1988 at the main cities in the Yellow River Basin and the NCP (Hebei Department of Water Conservancy, 1987-1988; Yellow River Conservancy Commission, 2002) was directly input to the model. In the 1990s, return flow was as much as 35% of withdrawal in the upper and 25% in the middle, but close to 0% in the downstream (Chen et al., 2003a; Cai and Rosegrant, 2004). The return flows at Qingtongxia and Hetao irrigation zones are 59% and 25% of withdrawal, whereas that at Weisan irrigation zone is close to 0%, because the river bed is above the level of the plain (Chen et al., 2003a; Cai and Rosegrant, 2004). This information was also input into the model.

At the upstream boundaries, a reflecting condition on the hydraulic head was used assuming that there is no inflow from the mountains in the opposite direction (Nakayama and Watanabe, 2004). At the eastern sea boundary, a constant head was set at 0 m. The hydraulic head values parallel to the observed ground level were input as initial conditions for the groundwater sub-model. As initial conditions, the ratios of river and aquifer irrigation were set at the same constant values as in the above section. In river grids decided by digital river network from 1:50,000 and 1:100,000 topographic maps (CAS, 1982), inflows or outflows from the riverbeds were simulated at each time step depending on the difference in the hydraulic heads of groundwater and river. The simulation area covered 3,000 km by 1,000 km with a grid spacing of 10 km, covering the entire Yellow River Basin and the NCP. The vertical layer was discretized in thickness with depth, with each layer increased in thickness by a factor of 1.1 (Nakayama, 2011b; Nakayama and Watanabe, 2008b; Nakayama et al., 2006). The upper layer was set at 2 m depth, and the 20th layer was defined as an elevation of –500 m from the sea surface. Simulations were performed with a time step of 6 h for two years during 1987–1988 after 6 months of warm-up period until equilibrium. The author first calibrated the simulated values including irrigation water use in 1987 against previous results, and then validated them in 1988. Previously observed data about river discharge (9 points; Yellow River Conservancy Commission, 1987-1988), soil moisture (7 points of the Global Soil Moisture Data Bank; Entin et al., 2000; Robock et al., 2000), and groundwater level (26 points; China Institute for Geo-Environmental Monitoring, 2003) were also used for the verification of the model (Fig. 1 and Table 2) in addition to values published in the literature (Clapp and Hornberger, 1978; Rawls et al., 1982). Details are described in Nakayama (2011b).

Reaches[a]	Irrigation area (x 10^4 ha)	
	GIS database (Liu 1996)	Previous research (Yang et al. 2004a)
Above LZ	46.8	39.5
LZ – TDG	342.1	344.1
TDG – LM	43.0	53.4
LM – SMX	295.6	281.3
SMX – HYK	59.2	60.6
Below HYK	160.7	155.0
Sum	947.3	933.9

[a]Abbreviation in the following; LZ, Lanzhou (R-1);
TDG, Toudaoguai (R-4); LM, Longmen; SMX, Sanmenxia;
HYK, Huayuankou (R-6).

Table 1. Comparison of irrigation area in the simulated condition with that in the previous research.

No.	Point Name	Type	Lat.	Lon.	Elev.(m)
R-1	Lanzhou	River Discharge	35°55.8'	103°19.8'	1794.0
R-2	Lanzhou	River Discharge	36°4.2'	103°49.2'	1622.0
R-3	Qingtongxia	River Discharge	38°21.0'	106°24.0'	1096.0
R-4	Toudaoguai	River Discharge	40°16.2'	111°4.2'	861.0
R-5	Hequ	River Discharge	39°22.2'	111°9.0'	861.0
R-6	Huayuankou	River Discharge	34°55.2'	113°39.0'	104.0
R-7	Lankao	River Discharge	34°55.2'	114°42.0'	73.0
R-8	Juancheng	River Discharge	35°55.8'	115°54.0'	1.0
R-9	Lijin	River Discharge	37°31.2'	118°18.0'	1.0
S-1	Bameng	Soil Moisture	40°46.2'	107°24.0'	1059.0
S-2	Xilingaole	Soil Moisture	39°4.8'	105°22.8'	1238.0
S-3	Yongning	Soil Moisture	38°15.0'	106°13.8'	1130.0
S-4	Xifengzhen	Soil Moisture	35°43.8'	107°37.8'	1435.0
S-5	Tianshui	Soil Moisture	34°34.8'	105°45.0'	1196.0
S-6	Lushi	Soil Moisture	34°0.0'	111°1.2'	675.0
S-7	Zhengzhou	Soil Moisture	34°49.2'	113°40.2'	99.0
G-1	Shanxi-1	Groundwater Level	37°44.0'	112°34.2'	772.51
G-2	Shanxi-2	Groundwater Level	38°0.7'	112°25.8'	831.10
G-3	Shanxi-3	Groundwater Level	37°58.0'	112°29.6'	788.96
G-4	Shanxi-4	Groundwater Level	37°47.3'	112°31.3'	779.03
G-5	Shanxi-5	Groundwater Level	37°43.2'	111°57.5'	780.51
G-6	Shanxi-6	Groundwater Level	40°3.4'	113°17.4'	1059.73
G-7	Shanxi-7	Groundwater Level	34°56.9'	110°45.1'	346.73
G-8	Shanxi-8	Groundwater Level	38°47.1'	112°44.0'	823.18
G-9	Shanxi-9	Groundwater Level	37°7.6'	111°54.2'	733.24
G-10	Henan-1	Groundwater Level	34°48.2'	114°18.2'	73.40
G-11	Henan-2	Groundwater Level	35°42.0'	115°1.3'	52.20
G-12	Henan-3	Groundwater Level	35°31.0'	115°1.0'	53.30
G-13	Henan-4	Groundwater Level	35°31.0'	115°12.5'	54.00
G-14	Henan-5	Groundwater Level	34°1.5'	113°50.8'	66.80
G-15	Henan-6	Groundwater Level	33°49.0'	113°56.3'	63.91
G-16	Henan-7	Groundwater Level	34°4.1'	115°18.2'	47.20
G-17	Henan-8	Groundwater Level	33°55.9'	116°22.3'	31.90
G-18	Henan-9	Groundwater Level	34°48.2'	114°18.2'	52.60
G-19	Qinghai-1	Groundwater Level	36°35.5'	101°44.7'	2321.17
G-20	Qinghai-2	Groundwater Level	37°0.0'	101°37.9'	2506.94
G-21	Qinghai-3	Groundwater Level	36°58.5'	101°38.9'	2474.63
G-22	Qinghai-4	Groundwater Level	36°34.3'	101°43.9'	2346.37
G-23	Qinghai-5	Groundwater Level	36°32.2'	101°40.5'	2451.32
G-24	Qinghai-6	Groundwater Level	36°43.1'	101°30.3'	2425.32
G-25	Qinghai-7	Groundwater Level	36°41.9'	101°30.9'	2383.98
G-26	Qinghai-8	Groundwater Level	36°14.8'	94°46.6'	3007.81

Table 2. Lists of observation stations for calibration and validation shown in Fig. 1.

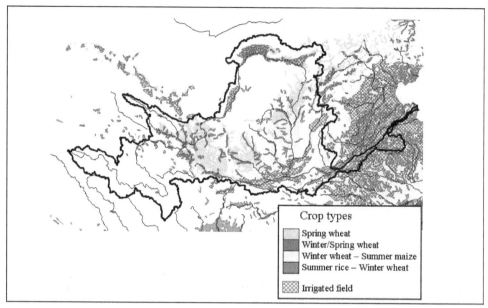

Fig. 3. Crop types in the agricultural areas. Irrigation areas are also overshaded. Irrigated fields cover most of the NCP for double cropping of winter wheat and summer maize in addition to three large irrigation zones.

3. Result and discussion

3.1 Verification of hydrologic cycle in the basin

The irrigation water use simulated in 1987 was firstly calibrated against previous results (Cai and Rosegrant, 2004; Liu and Xia, 2004; Yang et al., 2004a; Cai, 2006), showing a close agreement with the results of Cai and Rosegrant (2004). Then, the simulated value in 1988 was validated with the previous researches (Table 3), which indicates that there was reasonable agreement with each other and that irrigation water use is higher in large irrigation zones (LZ-TDG), along the Wei and Fen rivers (LM-SMX) in the middle, and in the downstream (below HYK). The results also show a high correlation between irrigation area and water use: r^2 = 0.986 (Chen et al., 2003a) and r^2 = 0.826 (Liu and Xia, 2004). Details of calibration and validation procedures are described in Nakayama (2011b).

The actual ET simulated by NICE reproduces reasonably the general trend estimated by integrated AVHRR NDVI data (Sun et al., 2004), which may give a good support on the predictive skill of the model (Fig. 4a-b). Although there are some discrepancies particularly for the lowest ET area (EP < 200 mm/year) mainly because of the banded colour figures, the simulated result reproduces the characteristics that the value is lowest in the downstream area of middle and on the Erdos Plateau—less than 200-300 mm per year (except in the irrigated area)—where vegetation is dominated by desert and soil is dominated by sand, and increases gradually towards the south-east. The simulated result also indicates that this spatial heterogeneity is related to human interventions and the resultant water stress by spring/winter cultivation in the upper/middle areas (Chen et al., 2003a; Tao et al., 2006), and winter wheat and summer maize cultivations in the middle/downstream (including the

Wei and Fen tributaries) and the NCP (Wang et al., 2001; Liu et al., 2002; Nakayama et al., 2006). Although the satellite-derived data are effective for grasping the spatial distribution of actual ET, there are some inefficiencies with regard to underestimation in sparsely vegetated regions (Inner Mongolia and Shaanxi Province) and overestimation in densely vegetated or irrigated regions (source area and Henan Province), as suggested by previous research (Sun et al., 2004; Zhou et al., 2007), which the simulation overcomes and improves mainly due to the inclusion of drought impact in the model. Details are described in Nakayama (2011b).

The model also simulated effect of irrigation on evapotranspiration at rotation between winter wheat and summer maize in the downstream of Yellow River (Fig. 4c). Because more water is withdrawn during winter-wheat period due to small rainfall in the north, the irrigation in this period affects greatly the increase in evapotranspiration. The simulated result indicates that the evapotranspiration increases predominantly during the seasons of grain filling and harvest of winter wheat with the effect of irrigation. In particular, most of the irrigation is withdrawn from aquifer in the NCP because surface water is seriously limited there (Nakayama, 2011b; Nakayama et al., 2006). This over-irrigation also affects the hydrologic change such as river discharge, soil moisture, and groundwater level in addition to evapotranspiration, as described in the following.

Reaches[a]	Irrigation water use (x 10^9 m³)				
	Simulated value (1988)[b]	Cai and Rosegrant 2004 (2000)[b]	Liu and Xia 2004 (1990s)[b]	Yang et al. 2004a (1990s)[b]	Cai 2006 (1988-1992)[b]
Above LZ	1.5	2.9	13.2		12.4
LZ – TDG	6.8	12.2			
TDG – LM	1.1	1.0		18.9	
LM – SMX	10.4	7.3	6.0		4.8
SMX – HYK	2.0	2.4			
Below HYK	8.4	10.6	10.8	9.5	11.2
Sum	30.2	36.4	30.0	28.5	28.4

[a]Abbreviation in the following; LZ, Lanzhou (R-1); TDG, Toudaoguai (R-4); LM, Longmen; SMX, Sanmenxia; HYK, Huayuankou (R-6).
[b]Value in parenthesis shows the target year in the simulation and the literatures.

Table 3. Validation of irrigation water use simulated by the model with that in the previous research.

The model could simulate reasonably the spatial distribution of irrigation water use after the comparison with a previous study based on the Penman-Monteith method and the crop coefficient (Fang et al., 2006) not only in reach level but also in the spatial distribution, as described in Nakayama (2011b). In particular, simulated ratios of river to total irrigation

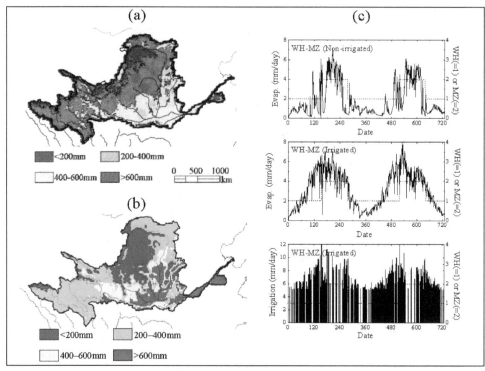

Fig. 4. Annual-averaged spatial distribution of evapotranspiration in 1987; (a) previous research; (b) simulated result; and (c) simulated value about impact of irrigation on evapotranspiration at rotation between winter wheat and summer maize. In Fig. 4c, right axis (dotted line) shows a period of each crop (WH; wheat, and MZ; maize).

(=river + aquifer) showed great variation and spatial heterogeneity in the basin. Fig. 5 shows the effect of over-irrigation on the decrease in river discharge on the downstream. The model reproduces reasonably the observed discharge for a low flow, and sometimes dry-up in the downstream (Yellow River Conservancy Commission, 1987-1988) with relatively high correlation r^2 and Nash-Sutcliffe criterion (NS; Nash and Sutcliffe, 1970) because the model includes the irrigation procedure and dynamic wave effect (Nakayama and Watanabe, 2004) in the model (Fig. 5b). The discharge decreases seriously in the downstream area mainly because of the water withdrawal for agriculture, which is more than 90% of the total withdrawal (Cai, 2006). At the downstream point at Lijin (R-9 in Table 2), the river discharge dries out during the spring mainly because most of the water is used for the irrigation of winter wheat in correspondence with the great increase in evapotranspiration shown in Fig. 4c. The model also indicated that the effect of groundwater irrigation is predominant in the downstream (data not shown), mainly on account of intensified water-use conflicts between upstream and downstream, and between various sectors like agriculture, municipality, and industry (Brown and Halweil, 1998; Nakayama, 2011a, 2011b; Nakayama et al., 2006). The smaller change in groundwater level in the upper was largely attributable to its unsuitability for crop production and the higher dependence of irrigation on surface water, as described previously (Yellow River Conservancy Commission, 2002).

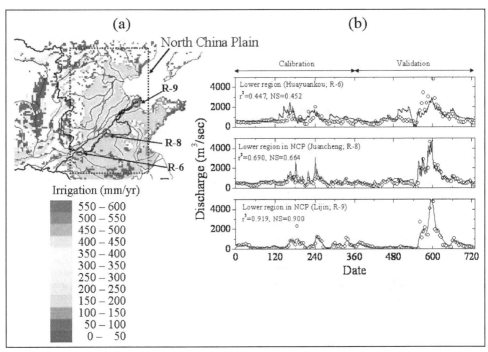

Fig. 5. Decrease in discharge caused by over-irrigation in the downstream region; (a) simulated result of river irrigation in 1987; (b) river discharge at the downstream. In Fig. 5b, solid line is the simulated result with irrigation, and circle is the observed value.

The simulated groundwater levels and soil moisture contents were calibrated and validated against observed data (Entin et al., 2000; Robock et al., 2000; China Institute for Geo-Environmental Monitoring, 2003) shown in Fig. 1 and Table 2 (data not shown). Although the correlation of groundwater level relative to the surface was not as good ($r^2 = 0.401$) as that of the absolute groundwater level ($r^2 = 0.983$) and the simulated value showed a tendency to overestimate the observed value in the calibration procedure for 1987, the simulation reproduced well the general distribution (BIAS = -21.2%, RMSE = 5.6 m, RRMSE = -0.468, MSSS = 0.356) (Nakayama, 2011b). This disagreement was due to the difference in surface elevation on the point-scale and mesh-scale (scale dependence), and the resolution of the groundwater flow model (changes in elevation from 0 m to 3000–4000 m in the basin). Because the simulated level is the hydraulic head in an aquifer, it might take a larger value than the land surface, particularly for a grid cell near or on the river. Another reason is that the irrigation water use simulated by the model might be underestimated because automatic irrigation supplied the water requirement for crops in order to satisfy the observed soil moisture, river discharge, groundwater level, LAI, evapotranspiration, and crop coefficient, which was theoretically pumped up from the river or the aquifer in the model. In reality, it has a possibility that farmers might use more irrigation water than the theoretical water requirement for crops if possible though there were not enough statistical or observed data to support it. The simulated water level decreases rapidly around the source area, indicating that there are many springs in this region. It is very low in the downstream (below sea level in some regions) because of the low elevation and overexploitation, as is the case in the NCP (Nakayama, 2011a, 2011b; Nakayama et al., 2006). The soil moisture is higher in the source area and in the paddy-dominated Qingtongxia Irrigation Zone (data not shown), corresponding closely with the distribution of the groundwater level. Details are described in Nakayama (2011b).

3.2 Impact of irrigation on hydrologic changes

Scenario analysis of conversion from unirrigated to irrigated run predicted the hydrologic changes (Fig. 6). The predicted result without irrigation generally overestimates the observed river discharge (Fig. 6a) and this effect is more prominent in the middle and downstream, as supported by reports that the difference between natural and observed runoff is larger downstream (Ren et al., 2002; Fu et al., 2004; Liu and Zheng, 2004). The difference between simulations considering and not considering irrigation strongly supports previous studies from the point of view that the influence of human interventions on river runoff has increased downstream over the last five decades (Chen et al., 2003a; Liu and Xia, 2004; Yang et al., 2004a; Cai, 2006; Tang et al., 2007) (Table 3), as also represented by the decline of water renewal times (Liu et al., 2003) and water resource renewability (Xia et al., 2004). This difference is greatly affected by complex irrigation procedures of various crops, which are roughly represented by spring/winter wheat in the upper-middle, and double cropping of winter wheat and summer maize in the middle-downstream regions (Wang et al., 2001; Liu et al., 2002; Fang et al., 2006; Nakayama et al., 2006; Tao et al., 2006).

Because there is some time lag between periods of increase in irrigation and decrease in runoff, the river discharge does not necessarily decrease in the winter and sometimes decreases in the summer. Further, the discharge sometimes increases slightly in the flood season, which indicates that the precipitation in irrigated fields sometimes responds quickly to flood drainage. Although both r^2 and NS have relatively low values across the basin (max: $r^2 = 0.447$, NS = 0.452), the simulated results with irrigation reproduce these characteristics better, and the statistics for MV (mean value), SD (standard deviation), and CV (coefficient of variation; CV = SD/MV) generally agree better with the observed values, as also supported by the better reproduction of other components of the hydrologic cycle, such as annual ET (Fig. 4) (data not shown in the case without irrigation) and irrigation water use (Fig. 5a, Table 3). The simulated result considering irrigation also reproduces the observed data for a low flow, and sometimes dry-up in the same way as Fig. 5b (Zhang et al., 1990; Yang et al., 1998; Ren et al., 2002), being attributable to inclusion of the dynamic wave effect in NICE, which other previous NICE series were unable to reproduce. Furthermore, the model improves the reproduction of river discharge in the basin in comparison with previous research (Yang and Musiake, 2003), where the ratio of absolute error to the mean was more than 60% at Huayuankou hydrological station, one of the worst such cases on a major river in Asia. The major reason for this disagreement is artificial water regulation such as reservoirs, water intakes, and diversions, which the model generally includes in addition to the extreme annual variation in flood seasons (Nakayama and Watanabe, 2008b).

Scenario analysis also predicts the groundwater level change and indicates that the effect of groundwater over-irrigation is predominant in the middle and downstream (Fig. 6b), where surface water is seriously limited, as shown in Fig. 5b and described in section 2.1 (Yellow River Conservancy Commission, 2002). The predicted result indicates a serious situation of water shortage in the downstream region and the NCP where groundwater level degrades over a wide area (Brown and Halweil). The result also implies that the model accounts for intensified water-use conflicts between upstream and downstream areas, and between agriculture, municipal, and industrial sectors (Brown and Halweil, 1998; Shimada, 2000; Chen et al., 2003b; Nakayama et al., 2006). These analyses of the impact of human intervention on hydrologic changes present strong indicatives of the seriousness of the situation, and imply the need for further correct estimation and appropriate measures against such irrigation loss and the low irrigation efficiency described previously (Wang et al., 2001). Details are described in Nakayama (2011b).

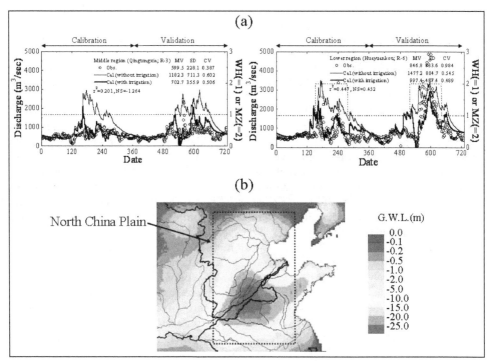

Fig. 6. Scenario analysis of conversion from unirrigated to irrigated run; (a) prediction of river discharge at the upper-middle (R–3; Qingtongxia) and the lower (R–6; Huayuankou) in Fig. 1 and Table 2; (b) groundwater level change in the middle-downstream regions. In Fig. 6a, circles show observation data; solid line is the simulated result without irrigation effect; bold line is the simulated result with irrigation. Right axis (dotted line) shows a period of each crop (WH; wheat, and MZ; maize) in the same way as Fig. 4c.

3.3 Discussion
Water scarcity and resource depletion in the downstream and the NCP, referred to as the 'bread basket' of China, is becoming more severe every year against increased crop production based on irrigation water, in addition to the expansion of municipal and industrial usage (Nakayama, 2011a; Nakayama et al., 2006). The simulated result shows the discharge was affected greatly by the rapid development of cities and industries and the increase in farmland irrigation (Fig. 6), which is closely related to severe groundwater degradation owing to the high clay content of the surface soil (Nakayama et al., 2006;

Nakayama, 2011a, 2011b). Because the dry-up of river reaches and groundwater exhaustion have been very severe so far (Chen et al., 2003b; Xia et al., 2004; Yang et al., 2004a), it is urgently necessary to perform effective control of water diversions (Liu and Xia, 2004; Liu and Zheng, 2004); the results simulated by NICE can be taken as strong indicatives of the seriousness of the situation. There are some reasons for the gap between irrigation water use (Fig. 5a) and groundwater level distribution (Fig. 6b). Firstly, the simulated levels have a temporally averaged distribution, and it takes some time for water levels to reach equilibrium after the boundary conditions have changed. This in turn affects the replenishment of groundwater from adjacent regions in addition to the heterogeneity of three-dimensional groundwater flow. Secondly, irrigation water is drawn not only from groundwater but also from river, and the ratio of river to total irrigation changes spatio-temporally in the basin (Fig. 5b); more river irrigation is drawn in the upper, and most of the irrigation depends on groundwater in the downstream, particularly in the NCP. This effect is clearly evident in comparison with the simulated results and the degradation value in the downstream is smaller that that in Fig. 6b.

Though the simulation reproduced reasonably hydrologic cycle such as evapotranspiration (Fig. 4), irrigation water use (Fig. 5a and Table 3), groundwater level, and river discharge (Fig. 5b and Fig. 6a), there were some discrepancies due to very complex and inaccurate nature of water withdrawal in the basin. In particular, the model achieved a relatively reasonable agreement though the model tried to calibrate and validate irrigation water use during only two years against other studies focusing on irrigation during long period (Fig. 5a and Table 3), which might lead to a substantial bias on model parameters. Because the objective of this study is primarily to evaluate the complex hydrological processes and reveal the impact of irrigation on hydrologic cycle in the basin through the verification during a fixed period, it is a future work to run model for the long period in the next step. At the same time, it will be of importance to derive better estimates of water demand in agricultural and urban areas during the long period by using more detailed statistical data, GIS data, and satellite data in longer period. Although the geological structure in the model included the general characteristics of several aquifers by reference to previous literature (Geological Atlas of China, 2002), the detailed structure of each aquifer layer was simplified as much as possible (Nakayama and Watanabe, 2008b; Nakayama et al., 2006). It will be necessary to obtain more precise data for the complex channel geometry of both natural and artificial rivers, soil properties, and geological structure. The spatial and temporal resolution used in the simulation also requires further improvement in order to overcome the problem of scale dependence and to improve verification and future reliability (Nakayama, 2011b).

Simulated results about the impact of irrigation on evapotranspiration change show a heterogeneous distribution (Fig. 4a-b). In particular, the irrigation of winter wheat increases greatly evapotranspiration, which is supplied by the limited water resources of river discharge and groundwater there (Fig. 5). This implies that energy supply is abundant relative to the water supply and the hydrological process is more sensitive to precipitation in the north, whereas the water supply is abundant relative to the energy supply and sun duration has a more significant impact in the south (Cong et al., 2010). The NICE is effective to provide better evaluation of hydrological trends in longer period including 'evaporation paradox' (Roderick and Farquhar, 2002; Cong et al., 2010) together with observation networks because the model does not need the crop coefficient (depending on a growing stage and a kind of crop) for the calculation of actual evaporation and simulates it directly without detailed site-specific information or empirical relation to calculate effective precipitation (Nakayama, 2011a; Nakayama et al., 2006). It is further necessary to clarify feedback and inter-relationship between micro, regional, and global scales; Linkage with global-scale dynamic vegetation model including two-way interactions between seasonal crop growth and atmospheric variability (Bondeau et al., 2007; Oleson et al., 2008); From stochastic to deterministic processes towards relationship between seedling establishment, mortality, and regeneration, and growth process based on carbon balance (Bugmann et al., 1996); From CERES-DSSAT to generic (hybrid) crop model by combinations of growth-development functions and mechanistic formulation of photosynthesis and respiration (Yang et al., 2004b); Improvement of nutrient fixation in seedlings, growth rate parameter, and stress factor, etc. for longer time-scale (Hendrickson et al., 1990). These future works might make a great contribution to the construction of powerful strategy for climate change problems in global scale.

Importance is that authority for water management in the basin is delineated by water source (surface water or groundwater) in addition to topographic boundaries (basin) and integrated water management concepts. In China, surface water and groundwater are managed by different authorities; the Ministry of Water Resources is responsible for surface water, while groundwater is considered a mineral resource and is administered by the Ministry of Minerals. In order to manage water resources effectively, any change in water accounting procedures may need to be negotiated through agreements brokered at relatively high levels of government, because surface water and groundwater are physically closely related to each other. Furthermore, the future development of irrigated and unirrigated fields and the associated crop production would affect greatly hydrologic change and usable irrigation water from river and aquifer, and vice versa (Nakayama, 2011b). The changes seen in this water resource are also related to climate change because groundwater storage moderates basin responses and climate feedback through evapotranspiration (Maxwell and Kollet, 2008). This is also related to a necessity of further evaluation about the evaporation paradox as described in the above. Although the groundwater level has decreased rapidly mainly due to overexploitation in the middle and downstream (Nakayama et al., 2006; Nakayama, 2011a, 2011b), regions where the land surface energy budget is very sensitive to groundwater storage are dominated by a critical water level (Kollet and Maxwell, 2008). The predicted hydrologic change indicates heterogeneous vulnerability of water resources and implies the associated impact on climate change (Fig. 6).

Basin responses will also be accelerated by an ambitious project to divert water from the Changjiang to the Yellow River, so-called, the South-to-North Water Transfer Project (SNWTP) (Rich, 1983; Yang and Zehnder, 2001). It can be estimated that the degradation of crop productivity may become severe, because most of the irrigation is dependent on vulnerable water resources (McVicar et al., 2002). Further research is necessary to examine the optimum amount of water that can be transferred, the effective management of the Three Gorges Dam (TGD) in the Changjiang River, the overall economic and social consequences of both projects, and their environmental assessment. It will be further necessary to obtain more observed and statistical data relating to water level, soil and water temperatures, water quality, and various phenological characteristics and crop productivity of spring/winter wheat and summer maize, in addition to satellite data of higher spatiotemporal resolution describing the seasonal and spatial vegetation phenology more accurately. The linear relationship between evapotranspiration and biomass production, which is very conservative and physiologically determined, is also valuable for further evaluation of the relationship between changes in water use and crop production by coupling with the numerical simulation and the satellite data analysis. Furthermore, it is powerful to develop a more realistic mechanism for sub-models, and to predict future hydrologic cycle and associated climate change using the model in order to achieve sustainable development under sound socio-economic conditions.

4. Conclusion

This study coupled National Integrated Catchment-based Eco-hydrology (NICE) model series with complex sub-models involving various factors, and clarified the importance of and diverse water system in the highly cultivated Yellow River Basin, including hydrological processes such as river dry-up, groundwater deterioration, agricultural water use, et al. The model includes different functions of representative crops (wheat, maize, soybean, and rice) and simulates automatically dynamic growth processes and biomass formulation. The model reproduced reasonably evapotranspiration, irrigation water use, groundwater level, and river discharge during spring/winter wheat and summer maize cultivations. Scenario analysis predicted the impact of irrigation on both surface water and groundwater, which had previously been difficult to evaluate. The simulated discharge with irrigation was improved in terms of mean value, standard deviation, and coefficient of variation. Because this region has experienced substantial river dry-up and groundwater degradation at the end of the 20th century, this approach would help to overcome substantial pressures of increasing food demand and declining water availability, and to decide on appropriate measures for whole water resources management to achieve sustainable development under sound socio-economic conditions.

5. Acknowledgment

The author thanks Dr. Y. Yang, Shijiazhuang Institute of Agricultural Modernization of the Chinese Academy of Sciences (CAS), China, and Dr. M. Watanabe, Keio University, Japan, for valuable comments about the study area. Some of the simulations in this study were run on an NEC SX-6 supercomputer at the Center for Global Environmental Research (CGER), NIES. The support of the Asia Pacific Environmental Innovation Strategy (APEIS) Project and the Environmental Technology Development Fund from the Japanese Ministry of Environment is also acknowledged.

6. References

Bondeau, A., Smith, P.C., Zaehle, S., Schaphoff, S., Lucht, W., Cramer, W., Gerten, D., Lotze-Campen, H., Muller, C., Reichstein, M. & Smith, B. (2007) Modelling the role of agriculture for the 20th century global terrestrial carbon balance. *Global Change Biol.*, Vol.13, pp.679-706, doi: 10.1111/j.1365-2486.2006.01305.x, ISSN 1354-1013
Brown, L.R. & Halweil, B. (1998). China's water shortage could shake world food security. *World Watch*, July/August, Vol.11(4), pp.10-18

Bugmann, H.K.M., Yan, X., Sykes, M.T., Martin, P., Linder, M., Desanker, P.V. & Cumming, S.G. (1996) A comparison of forest gap models: model structure and behaviour. *Climatic Change*, Vol.34, pp.289–313, ISSN 0165-0009

Cai, X. & Rosegrant, M.W. (2004). Optional water development strategies for the Yellow River Basin: Balancing agricultural and ecological water demands. *Water Resour. Res.*, Vol.40, W08S04, doi: 10.1029/2003WR002488, ISSN 0043-1397

Cai, X. (2006). *Water stress, water transfer and social equity in Northern China: Implication for policy reforms.* Human Development Report 2006, UNEP, Available from http://hdr.undp.org/en/reports/global/hdr2006/papers/cai ximing.pdf

Chen, J., He, D. & Cui, S. (2003a). The response of river water quality and quantity to the development of irrigated agriculture in the last 4 decades in the Yellow River Basin, China. *Water Resour. Res.*, Vol.39(3), 1047, doi: 10.1029/2001WR001234, ISSN 0043-1397

Chen, J.Y., Tang, C.Y., Shen, Y.J., Sakura, Y., Kondoh, A. & Shimada, J. (2003b). Use of water balance calculation and tritium to examine the dropdown of groundwater table in the piedmont of the North China Plain (NCP). *Environ. Geol.*, Vol.44, pp.564-571, ISSN 0943-0105

Chen, Y.M., Guo, G.S., Wang, G.X., Kang, S.Z., Luo, H.B. & Zhang, D.Z. (1995). *Main crop water requirement and irrigation of China.* Hydrologic and Electronic Press, Beijing, 73-102

China Institute for Geo-Environmental Monitoring (CIGEM) (2003). *China Geological Environment Infonet, Database of groundwater observation in the People's Republic of China,* Available from http://www.cigem.gov.cn

Chinese Academy of Sciences (CAS) (1982). *Topographic maps of 1:50,000 and 1:100,000*

Chinese Academy of Sciences (CAS) (1988). *Administrative division coding system of the People's Republic of China,* Beijing

Chinese Academy of Sciences (CAS) (2003). *China soil database,* Available from http://www.soil.csdb.cn

Clapp, R.B. & Hornberger, G.M. (1978). Empirical equations for some soil hydraulic properties. *Water Resour. Res.*, Vol.14, 601-604, ISSN 0043-1397

Cong, Z., Zhao, J., Yang, D. & Ni, G. (2010) Understanding the hydrological trends of river basins in China. *J. Hydrol.*, Vol.388, pp.350-356, doi: 10.1016/j.jhydrol.2010.05.013, ISSN 0022-1694

Doll, P. & Siebert, S. (2002). Global modeling of irrigation water requirements. *Water Resour. Res.*, Vol.38, 8-1—8-10, ISSN 0043-1397

Entin, J.K., Robock, A., Vinnikov, K.Y., Hollinger, S.E., Liu, S. & Namkhai, A. (2000). Temporal and spatial scales of observed soil moisture variations in the extratropics. *J. Geophys. Res.*, Vol.105(D9), pp.11865-11877, ISSN 0148-0227

Fang, W., Imura, H. & Shi, F. (2006). Wheat irrigation water requirement variability (2001-2030) in the Yellow River Basin under HADCM3 GCM scenarios. *Jpn. J. Environ. Sci.*, Vol.19(1), pp.3-14

Fu, G., Chen, S., Liu, C. & Shepard, D. (2004). Hydro-climatic trends of the Yellow River basin for the last 50 years. *Climatic Change*, Vol.65, pp.149-178, ISSN 0165-0009

Geological Atlas of China (2002). Geological Publisher, Beijing, China (in Chinese)

Godwin, D.C. & Jones, C.A. (1991). Nitrogen dynamics in soil-plant systems, In: *Modeling plant and soil systems*, Hanks, R.J. & Ritchie, J.T. (Eds.), 287-321, Agronomy 31, American Society of Agronomy, Madison, Wisconsin, USA

Hebei Department of Water Conservancy (1987). *Hebei year book of water conservancy for 1987* (in Chinese)

Hebei Department of Water Conservancy (1988). *Hebei year book of water conservancy for 1988* (in Chinese)

Hendrickson, O.Q., Fogal, W.H. & Burgess, D. (1990) Growth and resistance to herbivory in N2-fixing alders. *Can. J. Bot.*, Vol.69, pp.1919–1926, ISSN 0008-4026

Kollet, S.J., Maxwell, R.M., 2008. Capturing the influence of groundwater dynamics on land surface processes using an integrated, distributed watershed model. *Water Resour. Res.*, Vol.44, W02402, doi: 10.1029/2007WR006004, ISSN 0043-1397

Lee, T.M. (1996). Hydrogeologic controls on the groundwater interactions with an acidic lake in karst terrain, Lake Barco, Florida. *Water Resour. Res.*, Vol.32, 831-844, ISSN 0043-1397

Liu, C., Zhang, X. & Zhang, Y. (2002). Determination of daily evapotranspiration of winter wheat and corn by large-scale weighting lysimeter and micro-lysimeter. *Agr. Forest. Meteorol.*, Vol.111, pp.109-120, ISSN 0168-1923

Liu, C. & Xia, J. (2004). Water problems and hydrological research in the Yellow River and the Huai and Hai River basins of China. *Hydrol. Process.*, Vol.18, pp.2197-2210, doi: 10.1002/hyp.5524, ISSN 0885-6087

Liu, C. & Zheng, H. (2004). Changes in components of the hydrological cycle in the Yellow River basin during the second half of the 20th century. *Hydrol. Process.*, Vol.18, pp.2337-2345, doi: 10.1002/hyp.5534, ISSN 0885-6087

Liu, J.Y. (1996). *Macro-scale survey and dynamic study of natural resources and environment of China by remote sensing*, Chinese Science and Technology Publisher, Beijing, China (in Chinese)

Liu, L., Yang, Z. & Shen, Z. (2003). Estimation of water renewal times for the middle and lower sections of the Yellow River. *Hydrol. Process.*, Vol.17, pp.1941-1950, doi: 10.1002/hyp.1219, ISSN 0885-6087

Maxwell, R.M. & Kollet, S.J. (2008). Interdependence of groundwater dynamics and land-energy feedbacks under climate change. *Nat. Geosci.*, Vol.1, pp.665-669, doi: 10.1038/ngeo315, ISSN 1752-0894

McVicar, T.R., Zhang, G.L., Bradford, A.S., Wang, H.X., Dawes, W.R., Zhang, L. & Li, L. (2002). Monitoring regional agricultural water use efficiency for Hebei Province on the North China Plain. *Aust. J. Agric. Res.*, Vol.53, pp.55-76, ISSN 0004-9409

Nakayama, T. & Watanabe, M. (2004). Simulation of drying phenomena associated with vegetation change caused by invasion of alder (Alnus japonica) in Kushiro Mire. *Water Resour. Res.*, Vol.40, W08402, doi: 10.1029/2004WR003174, ISSN 0043-1397

Nakayama, T. & Watanabe, M. (2006). Simulation of spring snowmelt runoff by considering micro-topography and phase changes in soil layer. *Hydrol. Earth Syst. Sci. Discuss.*, Vol.3, pp.2101-2144, ISSN 1027-5606

Nakayama, T., Yang, Y., Watanabe, M. & Zhang, X. (2006). Simulation of groundwater dynamics in North China Plain by coupled hydrology and agricultural models. *Hydrol. Process.*, Vol.20(16), pp.3441-3466, doi: 10.1002/hyp.6142, ISSN 0885-6087

Nakayama, T., Watanabe, M., Tanji, K. & Morioka, T. (2007). Effect of underground urban structures on eutrophic coastal environment. *Sci. Total Environ.*, Vol.373(1), pp.270-288, doi: 10.1016/j.scitotenv.2006.11.033, ISSN 0048-9697

Nakayama, T. (2008a). Factors controlling vegetation succession in Kushiro Mire. *Ecol. Model.*, Vol.215, pp.225-236, doi: 10.1016/j.ecolmodel.2008.02.017, ISSN 0304-3800

Nakayama, T. (2008b). Shrinkage of shrub forest and recovery of mire ecosystem by river restoration in northern Japan. *Forest Ecol. Manag.*, Vol.256, pp.1927-1938, doi: 10.1016/j.foreco.2008.07.017, ISSN 0378-1127

Nakayama, T. & Watanabe, M. (2008a). Missing role of groundwater in water and nutrient cycles in the shallow eutrophic Lake Kasumigaura, Japan. *Hydrol. Process.*, Vol.22, pp.1150-1172, doi: 10.1002/hyp.6684, ISSN 0885-6087

Nakayama, T. & Watanabe, M. (2008b). Role of flood storage ability of lakes in the Changjiang River catchment. *Global Planet. Change*, Vol.63, pp.9-22, doi: 10.1016/j.gloplacha.2008.04.002, ISSN 0921-8181

Nakayama, T. & Watanabe, M. (2008c). Modelling the hydrologic cycle in a shallow eutrophic lake. *Verh. Internat. Verein. Limnol.*, Vol.30

Nakayama, T. (2009). Simulation of Ecosystem Degradation and its Application for Effective Policy-Making in Regional Scale, In: *River Pollution Research Progress*, Mattia N. Gallo & Marco H. Ferrari (Eds.), 1-89, Nova Science Publishers, Inc., ISBN 978-1-60456-643-7, New York

Nakayama, T. (2010). Simulation of hydrologic and geomorphic changes affecting a shrinking mire. *River Res. Appl.*, Vol.26(3), pp.305-321, doi: 10.1002/rra.1253, ISSN 1535-1459

Nakayama, T. & Fujita, T. (2010). Cooling effect of water-holding pavements made of new materials on water and heat budgets in urban areas. *Landscape Urban Plan.*, Vol.96, pp.57-67, doi: 10.1016/j.landurbplan.2010.02.003, ISSN 0169-2046

Nakayama, T., Sun, Y. & Geng, Y. (2010). Simulation of water resource and its relation to urban activity in Dalian City, Northern China. *Global Planet. Change*, Vol.73, pp.172-185, doi: 10.1016/j.gloplacha.2010.06.001, ISSN 0921-8181

Nakayama, T. (2011a). Simulation of complicated and diverse water system accompanied by human intervention in the North China Plain. *Hydrol. Process.*, Vol.25, pp.2679-2693 doi: 10.1002/hyp.8009, ISSN 0885-6087

Nakayama, T. (2011b). Simulation of the effect of irrigation on the hydrologic cycle in the highly cultivated Yellow River Basin. *Agr. Forest Meteorol.*, Vol.151, pp.314-327, doi: 10.1016/j.agrformet.2010.11.006, ISSN 0168-1923

Nakayama, T. & Hashimoto, S. (2011). Analysis of the ability of water resources to reduce the urban heat island in the Tokyo megalopolis. *Environ. Pollut.*, Vol.159, pp.2164-2173, doi: 10.1016/j.envpol.2010.11.016, ISSN 0269-7491

Nakayama, T., Hashimoto, S. & Hamano, H. (2011). Multi-scaled analysis of hydrothermal dynamics in Japanese megalopolis by using integrated approach. *Hydrol. Process.* (in press), ISSN 0885-6087

Nash, J.E. & Sutcliffe, J.V. (1970). Riverflow forecasting through conceptual model. *J. Hydrol.*, Vol.10, pp.282-290, ISSN 0022-1694

Oleson, K.W., Niu, G.-Y., Yang, Z.-L., Lawrence, D.M., Thornton, P.E., Lawrence, P.J., Stockli, R., Dickinson, R.E., Bonan, G.B., Levis, S., Dai, A. & Qian, T. (2008)

Improvements to the Community Land Model and their impact on the hydrological cycle. *J. Geophys. Res.*, Vol.113, G01021, doi: 10.1029/2007JG000563, ISSN 0148-0227

Oreskes, N., Shrader-Frechette, K. & Belitz, K. (1994). Verification, validation, and confirmation of numerical models in the earth sciences. *Science*, Vol.263, pp.641-646, ISSN 0036-8075

Priestley C.H.B. & Taylor, R.J. (1972). On the assessment of surface heat flux and evaporation using large-scale parameters. *Mon. Weather Rev.*, Vol.100, pp.81-92, ISSN 0027-0644

Rawls, W.J., Brakensiek, D.L. & Saxton, K.E. (1982). Estimation of soil water properties. *Trans. ASAE*, Vol.25, pp.1316-1320

Ren, L., Wang, M., Li, C. & Zhang, W. (2002). Impacts of human activity on river runoff in the northern area of China. *J. Hydrol.*, Vol.261, pp.204-217, ISSN 0022-1694

Rich, V. (1983). Yangtze to cross Yellow River. *Nature*, Vol.305, pp.568, ISSN 0028-0836

Ritchie, J.T., Singh, U., Godwin, D.C. & Bowen, W.T. (1998). Cereal growth, development and yield, In: *Understanding Options for Agricultural Production*, Tsuji, G.Y., Hoogenboom, G. & Thornton, P.K. (Eds.), 79-98, Kluwer, ISBN 0-7923-4833-8, Great Britain

Robock, A., Konstantin, Y.V., Govindarajulu, S., Jared, K.E., Steven, E.H., Nina, A.S., Suxia, L. & Namkhai, A. (2000). The global soil moisture data bank. *Bull. Am. Meteorol. Soc.*, Vol.81, pp.1281-1299, Available from
http://climate.envsci.rutgers.edu/soil_moisture/

Roderick, M.L. & Farquhar, G.D. (2002) The cause of decreased pan evaporation over the past 50 years. *Science*, Vol.298(15), pp.1410-1411, ISSN 0036-8075

Sato, Y., Ma, X., Xu, J., Matsuoka, M., Zheng, H., Liu, C. & Fukushima, Y. (2008). Analysis of long-term water balance in the source area of the Yellow River basin. *Hydrol. Process.*, Vol.22, pp.1618-1929, doi: 10.1002/hyp.6730, ISSN 0885-6087

Sellers, P.J., Randall, D.A., Collatz, G.J., Berry, J.A., Field, C.B., Dazlich, D.A., Zhang, C., Collelo, G.D. & Bounoua, L. (1996). A revised land surface prameterization (SiB2) for atomospheric GCMs. Part I : Model formulation. *J. Climate*, Vol.9, pp.676-705, ISSN 0894-8755

Shimada, J. (2000). Proposals for the groundwater preservation toward 21st century through the view point of hydrological cycle. *J. Japan Assoc. Hydrol. Sci.*, Vol.30, pp.63-72 (in Japanese)

Sun, R., Gao, X., Liu, C.M. & Li, X.W. (2004). Evapotranspiration estimation in the Yellow River Basin, China using integrated NDVI data. *Int. J. Remote Sens.*, Vol.25, pp.2523-2534, ISSN 0143-1161

Tang, Q., Oki, T., Kanae, S. & Hu, H. (2007). The influence of precipitation variability and partial irrigation within grid cells on a hydrological simulation. *J. Hydrometeorol.*, Vol.8, pp.499-512, doi: 10.1175/JHM589.1, ISSN 1525-755X

Tang, Q., Oki, T., Kanae, S. & Hu, H. (2008a). Hydrological cycles change in the Yellow River basin during the last half of the twentieth century. *J. Climate*, Vol.21, pp.1790-1806, doi: 10.1175/2007JCLI1854.1, ISSN 0894-8755

Tang, Q., Oki, T., Kanae, S. & Hu, H. (2008b). A spatial analysis of hydro-climatic and vegetation condition trends in the Yellow River basin. *Hydrol. Process.*, Vol.22, pp.451-458, doi: 10.1002/hyp.6624, ISSN 0885-6087

Tao, F., Yokozawa, M., Xu, Y., Hayashi, Y. & Zhang, Z. (2006). Climate changes and trends in phenology and yields of field crops in China, 1981-2000. *Agr. Forest Meteorol.*, Vol.138, pp.82-92, ISSN 0168-1923

U.S. Geological Survey (USGS) (1996). *GTOPO30 Global 30 Arc Second Elevation Data Set*, USGS, Available from http://www1.gsi.go.jp/geowww/globalmap-gsi/gtopo30/gtopo30.html

Wang, H., Zhang, L., Dawes, W.R. & Liu, C. (2001). Improving water use efficiency of irrigated crops in the North China Plain – measurements and modeling. *Agr. Forest. Meteorol.*, Vol.48, pp.151-167, ISSN 0168-1923

Xia, J., Wang, Z., Wang, G. & Tan, G. (2004). The renewability of water resources and its quantification in the Yellow River basin, China. *Hydrol. Process.*, Vol.18, pp.2327-2336, doi: 10.1002/hyp.5532, ISSN 0885-6087

Xu, Z.X., Takeuchi, K., Ishidaira, H. & Zhang, X.W. (2002). Sustainability analysis for Yellow River Water Resources using the system dynamics approach. *Water Resour. Manag.*, Vol.16, pp.239-261, ISSN 0920-4741

Yang, Z.S., Milliman, J.D., Galler, J., Liu, J.P. & Sun, X.G. (1998). Yellow River's water and sediment discharge decreasing steadily. *EOS*, Vol.79(48), pp.589-592, ISSN 0096-3941

Yang, H. & Zehnder, A. (2001). China's regional water scarcity and implications for grain supply and trade. *Environ. Plann. A*, Vol.33, pp.79-95

Yang, D. & Musiake, K. (2003). A continental scale hydrological model using the distributed approach and its application to Asia. *Hydrol. Process.*, Vol.17, pp.2855-2869, doi: 10.1002/hyp.1438, ISSN 0885-6087

Yang, D., Li, C., Hu, H., Lei, Z., Yang, S., Kusuda, T., Koike, T. & Musiake, K. (2004a). Analysis of water resources variability in the Yellow River of China during the last half century using historical data. *Water Resour. Res.*, Vol.40, W06502, doi: 10.1029/2003WR002763, ISSN 0043-1397

Yang, H.S., Dobermann, A., Lindquist, J.L., Walters, D.T., Arkebauer, T.J. & Cassman, K.G. (2004b) Hybrid-maize–a maize simulation model that combines two crop modeling approaches. *Field Crop. Res.*, Vol.87, pp.131-154, ISSN 0378-4290

Yellow River Conservancy Commission (1987). *Annual report of discharge and sediment in Yellow River*, Interior report of the committee (in Chinese)

Yellow River Conservancy Commission (1988). *Annual report of discharge and sediment in Yellow River*, Interior report of the committee (in Chinese)

Yellow River Conservancy Commission (2002). *Yellow River water resources bulletins*, Available from http://www.yrcc.gov.cn/ (in Chinese)

Zhang, J., Huang, W.W. & Shi, M.C. (1990). Huanghe (Yellow River) and its estuary: sediment origin, transport and deposition. *J. Hydrol.*, Vol.120, pp.203-223, ISSN 0022-1694

Zhou, M.C., Ishidaira, H. & Takeuchi, K. (2007). Estimation of potential evapotranspiration over the Yellow River basin: reference crop evaporation or Shuttleworth-Wallance?. *Hydrol. Process.*, Vol.21, pp.1860-1874, doi: 10.1002/hyp.6339, ISSN 0885-6087

Zhu, Y. (1992). *Comprehensive hydro-geological evaluation of the Huang-Huai-Hai Plain*, Geological Publishing House of China, 277p., Beijing, China (in Chinese)

7

Using Soil Moisture Data to Estimate Evapotranspiration and Development of a Physically Based Root Water Uptake Model

Nirjhar Shah[1], Mark Ross[2] and Ken Trout[2]

[1]*AMEC Inc. Lakeland, FL*
[2]*Univ. of South Florida, Tampa, FL*
USA

1. Introduction

In humid regions such as west-central Florida, evapotranspiration (ET) is estimated to be 70% of precipitation on an average annual basis (Bidlake et al. 1993; Knowles 1996; Sumner 2001). ET is traditionally inferred from values of potential ET (PET) or reference ET (Doorenabos and Pruitt 1977). PET data are more readily available and can be computed from either pan evaporation or from energy budget methods (Penman 1948; Thornthwaite 1948; Monteith 1965; Priestly and Taylor 1972, etc.). The above methodology though simple, suffer from the fact that meteorological data collected in the field for PET are mostly under non-potential conditions, rendering ET estimates as erroneous (Brutsaert 1982; Sumner 2006). Lysimeters can be used to determine ET from mass balance, however, for shallow water table environments, they are found to give erroneous readings due to air entrapment (Fayer and Hillel 1986), as well as fluctuating water table (Yang et al. 2000). Remote sensing techniques such as, satellite-derived feedback model and Surface Energy Balance Algorithm (SEBAL) as reviewed by Kite and Droogers (2000) and remotely sensed Normalized Difference Vegetation Index (NDVI) as used by Mo et al. (2004) are especially useful for large scale studies. However, in the case of highly heterogeneous landscapes , the resolution of ET may become problematic owing to the coarse resolution of the data (Nachabe et al. 2005). The energy budget or eddy correlation methodologies are also limited to computing net ET and cannot resolve ET contribution from different sources. For shallow water table environments, continuous soil moisture measurements and water table estimation have been found to accurately determine ET (Nachabe et al. 2005; Fares and Alva 2000). Past studies, e.g., Robock et al. (2000), Mahmood and Hubbard (2003), and Nachabe et al. (2005), have clearly shown that soil moisture monitoring can be successfully used to determine ET from a hydrologic balance. The approach used herein involves use of soil moisture and water table data measurements. Using point measurement of soil moisture and water table observations from an individual monitoring well ET values can be accurately determined. Additionally, if similar measurements of soil moisture content and water table are available from a set of wells along a flow transect , other components of water budgets and attempts to comprehensively resolve other components of the water budget at the study site.

The following section describes a particular configuration of the instruments, development of a methodology, and an example case study where the authors have successfully applied measurement of soil moisture and water table in the past to estimate and model ET at the study site. The authors also used the soil moisture dataset to compute actual root water uptake for two different land-covers (grassed and forested). The new methodology of estimating ET is based on an eco-hydrological framework that includes plant physiological characteristics. The new methodology is shown to provide a much better representation of the ET process with varying antecedent conditions for a given land-cover as compared to traditional hydrological models.

2. Study site

The study site for gathering field data and using it for ET estimation and vadose zone process modeling was located in the sub basin of Long Flat Creek, a tributary of the Alafia River, adjacent to the Tampa bay regional reservoir, in Lithia, Florida. **Figure 1** shows the regional and aerial view of the site location. Two sets of monitoring well transects were installed on the west side of Long Flat Creek. One set of wells designated as PS-39, PS-40, PS-41, PS-42, and PS-43 ran from east to west while the other set consisting of two wells was roughly parallel to the stream (Long Flat Creek), running in the North South direction. The wells were designated as USF-1 and USF-3.

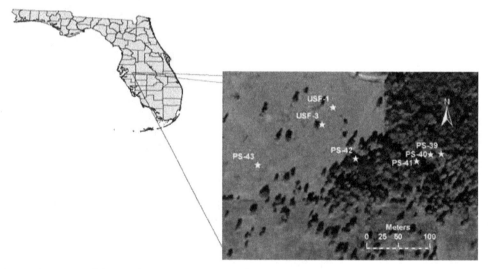

Fig. 1. Location of the study site in Hillsborough County, Florida

The topography of the area slopes towards the stream with PS-43 being located at roughly the highest point for both transects. The vegetation varied from un-grazed Bahia pasture grass in the upland areas (in proximity of PS-43, USF-1, and USF-3), to alluvial wetland forest comprised of slash pine and hardwood trees near the stream. The area close to PS-42 is characterized as a mixed (grassed and forested) zone. Horizontal distance between the wells is approximately 16, 22, 96, 153 m from PS-39 to PS-43, with PS-39 being approximately 6 m from the creek. The horizontal distance between USF-1 and USF-3 was 33 m. All wells were surveyed and land surface elevations were determined with respect to National Geodetic Vertical Datum 1929 (NGVD).

The data captured from this configuration was used both for point estimation as well as transect modeling, however , for this particular chapter, only point estimation of ET and and point data set will be used to develop conceptualizations of vadose zone processes will be discussed. For details regarding transect modeling to generate water budget estimates refer to Shah (2007).

3. Instrumentation

For measurement of water table at a particular location a monitoring well instrumented with submersible pressure transducer (manufactured by Instrumentation Northwest, Kirkland, WA) 0-34 kPa (0-5 psi), accurate to 0.034 kPa (0.005 psi) was installed. Adjacent to each well, an EnviroSMART® soil moisture probe (Sentek Pty. Ltd., Adelaide, Australia) carrying eight sensors was installed (see **Figure 2**). The soil moisture sensors allowed measurement of volumetric moisture content along a vertical profile at different depths from land surface. The sensors were deployed at 10, 20, 30, 50, 70, 90, 110, 150 cm from the land surface. The sensors work on the principle of frequency domain reflectometery (FDR) to convert electrical capacitance shift to volumetric water content ranging from oven dryness to saturation with a resolution of 0.1% (Buss 1993). Default factory calibration equations were used for calibrating these sensors. Fares and Alva (2000) and Morgan et al. (1999) found no significant difference in the values of observed recorded water content from the sensors when compared with the manually measured values. Two tipping bucket and two manual rain gages were also installed to record the amount of precipitation.

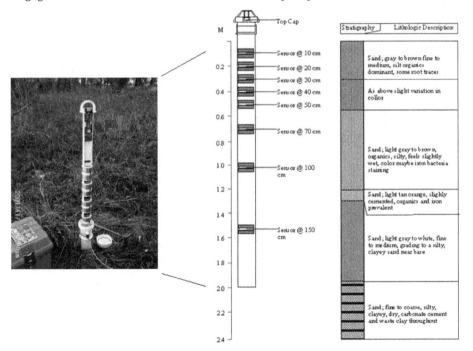

Fig. 2. Soil moisture probe on the left showing the mounted sensors along with schematics on the right showing sample stratiagraphy at different depths.

4. Point estimation of evapotranspiration using soil moisture data

At any given well location variation in total soil moisture on non-rainy days can be due to (a) subsurface flow from or to the one dimensional soil column (0 – 155 cm below land surface) over which soil moisture is measured, and (b) evapotranspiration from this soil column. Mathematically

$$\frac{\partial TSM}{\partial t} = Q - ET \tag{1}$$

where t is time [T], Q is subsurface flow rate [LT^{-1}], and ET is evapotranspiration rate [LT^{-1}]. TSM is total soil moisture, determined as below

$$TSM = \int_{\varsigma} \theta \, dz \tag{2}$$

where $\theta[L^3L^{-3}]$ is the measured water content, z [L] is the depth below land surface ς[L] is the depth of monitored soil column (155 cm). The values in the square brackets (for all the variables) represent the dimensions (instead of units) e.g. L is length, T is time.

The negative sign in front of ET in **Equation 1** indicates that ET depletes the TSM in the column. The subsurface flow rate can be either positive or negative. In a groundwater discharge area, the subsurface flow rate, Q, is positive because it acts to replenish the TSM in the soil column (Freeze and Cherry, 1979). Thus, this flow rate is negative in a groundwater recharge area. **Figure 3** illustrates the role of subsurface flow in replenishing or depleting total soil moisture in the column. An inherent assumption in this approach is that the deepest sensor is below the water table which allows accounting for all the soil moisture in the vadose zone. Hence, monitoring of water table is critical to make sure that the water table is shallower than the bottom most sensor. To estimate both ET and Q in **Equation 1**, it was important to decouple these fluxes. In this model the subsurface flow rate was estimated from the diurnal fluctuation in TSM. Assuming ET is effectively zero between midnight and 0400 h, Q can be easily calculated from **Equation 3** using:

$$Q = \frac{TSM_{0400h} - TSM_{midnight}}{4} \tag{3}$$

where TSM_{0400h} and $TSM_{midnight}$ are total soil moisture measured at 0400 h and midnight, respectively. The denominator in **Equation 3** is 4 h, corresponding to the time difference between the two TSM measurements. The assumption of negligible ET between midnight and 0400h is not new, but was adopted in the early works of White (1932) and Meyboom (1967) in analyzing diurnal water table fluctuation. It is a reasonable assumption to make at night when sunlight is absent.

Taking Q as constant for a 24h period (White 1932; Meyboom, 1967), the ET consumption in any single day was calculated from the following equation

$$ET = TSM_j - TSM_{j+1} + 24 \times Q \tag{4}$$

where TSM_j is the total soil moisture at midnight on day j, and TSM_{j+1} is the total soil moisture 24h later (midnight the following day). Q is multiplied by 24 as the **Equation 4** provides daily ET values. **Figure 4** show a sample observations for 5 day period showing the evolution of TSM in a groundwater discharge and recharge area respectively. Also marked on the graphs are different quantities calculated to determine *ET* from the observations.

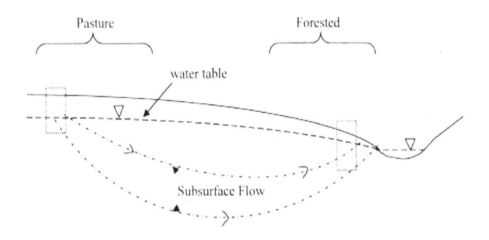

Fig. 3. Total soil moisture is estimated in two soil columns. The first is in a groundwater recharge area (pasture), and the second is in a groundwater discharge area (forested). In a groundwater discharge area, subsurface flow acts to replenish the total soil.

Equation 1 applies for dry periods only, because it does not account for the contribution of interception storage to ET on rainy days. Also, the changes in soil moisture on rainy days can occur due to other processes like infiltration, upstream runoff infiltration (as will be discussed later) etc. The results obtained from the above model were averaged based on the land cover of each well and are presented as ET values for grass or forested land cover. The values for the grassed land cover were also compared against ET values derived from pan evaporation measurements.

The ET estimates from the data collected at the study site using the above methodology are shown in **Figure 5**. **Figure 5** shows variability in the values of *ET* for a period of about a year and half. It can be seen from **Figure 5** that the method was successful in capturing spatial variability in the ET rates based on the changes in the land cover, as the ET rate of forested (alluvial wetland forest) land cover was found to be always higher than that of the grassland (in this case un-grazed Bahia grass). In addition to spatial variability, the method seemed to capture well the temporal variability in ET. The temporal variability for this particular analysis existed at two time scales, a short-term daily variation associated with daily changes in atmospheric conditions (e.g. local cloud cover, wind speed etc.) and a long-term, seasonal, climatic variation. The short-term variation tends to be less systematic and is demonstrated in **Figure 5** by the range marks. The seasonal variation is more systematic and pronounced and is clearly captured by the method.

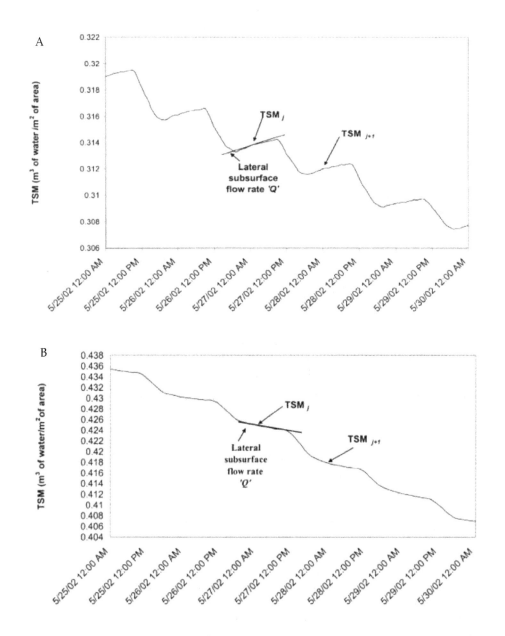

Fig. 4. Total soil moisture versus time in the (a) groundwater discharge area and (b) ground water recharge area. The subsurface flux is the positive slope of the line between midnight and 4 AM.

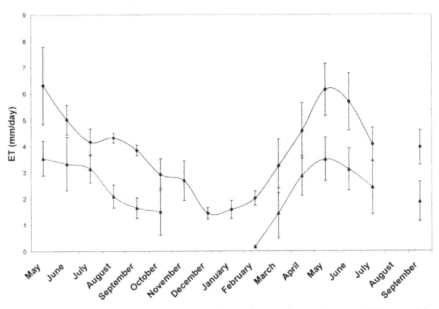

Fig. 5. Monthly average of evapotranspiration (ET) daily values in forested (diamonds) and pasture (triangles) areas. The gap in the graph represents a period of missing data. Standard deviations of daily values are also shown in the range limits.

To assess the reasonableness of the methodology, the estimated ET values for pasture were compared with ET estimated from the evaporation pan. The measured pan evaporation was multiplied by a pan coefficient for pasture to estimate ET for this vegetation cover. A monthly variable crop coefficient was adopted (Doorenbos and Pruitt, 1977) to account for changes associated with seasonal plant phenology (see **Table 1**). The consumptive water use or the crop evapotranspiration is calculated as:

$$ET_C = E_P \times K_C \qquad (5)$$

where E_P is the measured pan evaporation, K_C is a pan coefficient for pastureland, and ET_C is the estimated evapotranspiration [LT^{-1}] (mm/d) by the pan evaporation method. **Figure 6** compares the ET estimated by both the evaporation pan and moisture sensors for pasture. Although the two methods are fundamentally different, on average, estimated ET agreed well with an R^2 coefficient of 0.78. This supported the validity of the soil moisture methodology, which further captured the daily variability of ET ranging from a low of 0.3 mm/d to a maximum of 4.9 mm/d. The differences between the two methods can be attributed to fundamental discrepancies. The pan results are based on atmospheric potential with crude average monthly coefficients while the TSM approach inherently incorporates plant physiology and actual moisture limitations. Indeed, both methods suffer from limitations. The pan coefficient is generic and does not account for regional variation in vegetation phenology or other local influences such as soil texture and fertility. Similarly, the accuracy of the soil moisture method proposed in this study depends on the number of sensors used in monitoring total moisture in the soil column.

Month		Coefficient
January		0.4
February		0.45
March		0.55
April		0.64
May		0.7
June		0.7
July		0.7
August		0.7
September		0.7
October		0.6
November		0.5
December		0.5

Table 1. Pan coefficients used to obtain pasture evapotranspiration for different months.

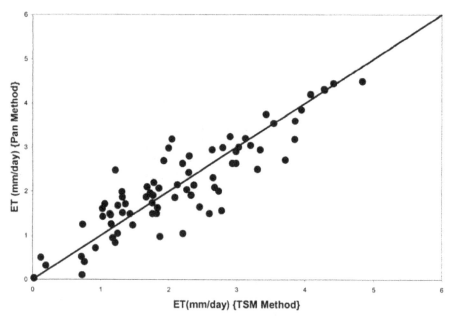

Fig. 6. Evapotranspiration estimates for pasture by the pan and point scale model. Data points represent the daily values of ET from both techniques.

5. Development of root water uptake model

The preceding sections described a novel data collection approach that can be used to measure ET (and other water budget components). The measured values can be subsequently used to develop modeling parameters or validate modeling results for areas which are similar to the study site in terms of climatic and land-cover conditions. The next step is the development of a generic modeling framework to accurately determine ET.

Transpiration by its very nature is a process that is primarily based on plant physiology and the better one can determine root water uptake the more accurate will be the estimation of actual transpiration and, therefore ET. Traditionally used models and concepts, however, make over simplifying assumptions about plants (Shah et al. 2007), hence casting doubt on the model results. What needs to be done is to try and combine land cover characteristics in the root water uptake models to produce more reliable results. With this intent in mind, recently, a new branch of study called "Eco-Hydrology" has been initiated. The aim of eco-hydrology is to encourage the interdisciplinary work on ecology and hydrology with an objective of improving hydrological modeling capabilities.

Soil moisture datasets (as described in Section 2) can be used to provide insight into the process of root water uptake which can then be combined with plant characteristics to develop a more physically based ET model. The next sections describe how the soil moisture dataset has been used by the authors to estimate vertical distribution of root-water uptake for two land-cover classes (shallow rooted and deep rooted) and how the results were then used to develop a land-cover based modeling framework.

5.1 Traditional root water uptake models

The governing equation for soil moisture dynamics in the unsaturated soil zone is the Richards's equation (Richards 1931). Richards's equation is derived from Darcy's law and the continuity equation. What follows is a brief description of Richards's equation and how can it incorporates root water uptake. For more detailed information about formulation of Richards's equation, including its derivation in three dimensions, the readers are directed to any text book on soil physics e.g. Hillel (1998).

Due to ease of measurement and conceptualization, energy of water (E) is represented in terms of height of liquid column and is called the hydraulic head (h). It is defined as the total energy of water per unit weight. Mathematically hydraulic head, h, can be represented as

$$h = \frac{E}{\rho_W g} \tag{6}$$

where ρ_W is the density of water and g is the acceleration due to gravity. The flow of water always occurs along decreasing head. In soil physics the fundamental equation used to model the flow of water along a head gradient is known as Darcy Law (Hillel 1998). Mathematically the equation can be written as

$$q = K \frac{\Delta h}{l} \tag{7}$$

where q [$L^3 L^{-2} T^{-1}$] is known as the specific discharge and is defined as the flow per unit cross-sectional area, K[LT^{-1}] is termed as the hydraulic conductivity, which indicates ease of flow, Δh [L] is the head difference between the points of interest and l[L] is the distance between them. Darcy's Law is analogous to Ohm's law with head gradient being analogous to the potential difference and current being analogous to specific discharge and hydraulic conductivity being similar to the conductance of a wire.

The second component of Richards's equation is the continuity equation. Continuity equation is based on the law of mass conservation, and for any given volume it states that net increase in storage in the given volume is inflow minus sum of outflow and any sink present in the volume of soil. Mathematically, it is this sink term that allows the modeling of water extracted from the given volume of soil.

In one dimension for flow occurring in the vertical direction (z axis is positive downwards) Richards's equation can be written as

$$\left(\frac{\partial \theta}{\partial t}\right) = \left(\frac{\partial}{\partial z} K \left(\frac{\partial h}{\partial z}+1\right)\right) - S \tag{8}$$

where θ is the water content, defined as the ratio of volume of water present and total volume of the soil element , t is time, S represents the sink term while other terms are as defined before.

If flow in X and Y directions is also considered , Richards's equation in three dimensions can be derived. Solution of Equation 8 can theoretically provide the spatial and temporal variability of moisture in the soil. However, due to high degree of non linearity of the equation no analytical solution exists for Richards's equation and numerical techniques are used to solve it. For a numerical solution of Richards's equation two essential properties that need to be defined a-priori are (a) relationship between soil water content and hydraulic head, also known as, soil moisture retention curves, and (b) a model that relate hydraulic head to root water uptake. Details about the soil moisture retention curves and numerical techniques used to solve Richards's equation can be found in Simunek et al. (2005). While much literature and field data exist describing the soil moisture retention curves, relatively less information exists about root water uptake models. The root water uptake models generally used, especially, on a watershed scale, are mostly empirical and lack any field verification.

The most common approach used to model root water uptake is to define a sink term S as a function of hydraulic head using the following equation

$$S(h) = \alpha(h) S_p \tag{9}$$

where $S(h)[L^3L^{-3}T^{-1}]$ is the actual root water uptake (RWU) from roots subjected to hydraulic or capillary pressure head 'h'. On the right hand side of the equation S_p $[L^3L^{-3}T^{-1}]$ is the maximum (also known as potential) uptake of water by the roots. The $a(h)$ is a root water uptake stress response function, with its values varying between 0 and 1.

The idea behind conceptualization of **Equation 9** is based on three basic assumptions. The first assumption being , as the soil becomes dryer the amount of water that can be extracted will decrease proportionally. Secondly, the amount of water extracted by the roots is affected by the ambient climatic conditions. Drier and hotter conditions result in more water loss from surface of leaves, hence, initiating more water extraction from the soil. The third and final assumption is that the uptake of water from a particular section of a root is directly proportional to the amount of roots present in that section.

The root water stress response function (α) is a result of the first assumption. Two models commonly used to define α are the Feddes model (Feddes et al. 1978) and the van Genuchten model (van Genuchten 1987). **Figure 7** (a and b, respectively) show the variation of α with decreasing hydraulic head which is same as decreasing water content or increasing soil dryness. Both models for α are empirical and do not involve any plant physiology to define the thresholds for the water stress response function. An interesting contrast, due to empiricism that is clearly evident is the value of α during saturated conditions. While the Feddes model predict the value of a to decrease to zero van Genuchten model predicts totally opposite with α rising to become unity under saturated conditions.

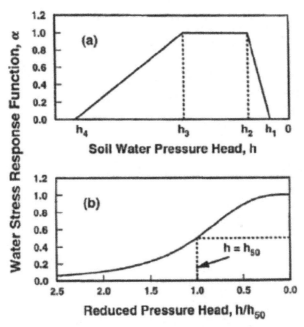

Fig. 7. Water stress response function as given conceptualized by (a) Feddes et al. 1978 and (b) van Genuchten (1980) [Adapted from Simunek et al. 2005].

Recently couple of different models (Li et al. 2001, Li et al. 2006) have been presented to overcome the empiricism in α. However these models are more a result of observation fitting and fail to bring in the plant physiology, which is what causes the changes in the water uptake rate due variation in soil moisture conditions.

Combining the second and the third assumptions in **Equation 9** results in the definition of S_p. S_p for any section of roots is defined as the product of root fraction in that section and the maximum possible water loss by the plant which is also known as the potential evapotranspiration. Potential evapotranspiration is a function of ambient atmospheric conditions and standard models like Penman-Monteith (Allen et al. 1998) are used to calculate the potential evapotranspiration rate. The problem with this definition of Sp is that for any given value of potential evapotranspiration, limiting the value of root water uptake by the root-fraction restrict the amount of water that can be extracted from a particular section. In other words, the amount of water extracted by a particular section of root is directly proportional to the amount of roots present and ignores the amount of ambient soil moisture present. This as will discussed later using field data is a significant limitation especially during dry period when the top soil with maximum roots get dry while the deep soil layer with lesser root mass still have soil moisture available for extraction.

5.2 Use of soil moisture data to estimate root water uptake

For the current analysis, the soil moisture data as described in Section 2 are used. Soil moisture and water-table data from well locations PS-43 and PS-40 were used to determine root water uptake from forested versus grassed land cover. The well PS-43 is referred to as Site A while PS-40 will be called Site B. Hourly averaged data at four hour time step were used for the analysis.

Extensive soil investigations including in-situ and laboratory analysis were performed for the study site. The soil in the study area is primarily sandy marine sediments with high permeability in the surface and subsurface layers. Detailed information about soil and site characteristics can be found in Said et al. (2005), and Trout and Ross (2004). Data for period of record January 2003 to December 2003 were used in this analysis.

van Genuchten (1980) proposed a model relating the water content and hydraulic conductivity with the suction head (soil suction pressure) represented by the following equations

$$S_e = \frac{\theta - \theta_r}{\theta_s - \theta_r} \tag{10}$$

$$h(\theta) = \frac{(S_e^{\frac{1}{m}} - 1)^{\frac{1}{n}}}{\phi} \tag{11}$$

$$K(h) = \begin{cases} K_S S_e^l [1 - (1 - S_e^{1/m})^m] & h < 0 \\ K_S & h \geq 0 \end{cases} \tag{12}$$

where m = 1 - 1/n for n > 1, S_e [-] is the normalized water content, varying between 0 and 1. θ is the observed water content, while θ_r and θ_s are the residual and saturated water content values respectively K_S [LT^{-1}] is the hydraulic conductivity when the soil matrix is saturated, l[-] is the pore connectivity parameter assumed to be 0.5 as an average for most soils (Mualem, 1976), and ϕ[L^{-1}], n[-] and m[-] are the van Genuchten empirical parameters. Negative values of hydraulic head (suction head) indicate the water content in the soil matrix is less than saturated while the positive value indicate saturated conditions. From the **Equations 11** and **12**, it is clear that for each type of soil five parameters, namely, K_S, n, ϕ, θ_r and θ_s have to be determined to uniquely define relationship of hydraulic conductivity and water content with soil suction head.

Figure 8 shows the schematics of the vertical soil column which is monitored using eight soil moisture sensors and a pressure transducer measuring the water table elevation, at each of the two locations. Shown also in **Figure 8** is the zone of influence of each sensor along with the elevation of water table and arrows showing possible flow directions. For the purpose of defining moisture retention and hydraulic conductivity curves, each section is treated as a different soil layer and independently parameterized. Hence, for each of the two locations for this particular study eight soil cores from depths corresponding to the zone of influence of each sensor were taken and analyzed (see Shah, 2007 for more details). **Table 2(a)** and **(b)** shows the parameters values that were obtained following the all the soil tests.

Sensor Location Below Land Surface (cm)	$\theta_S(\%)$	$\theta_R(\%)$	Φ (cm^{-1})	n(-)	K_S(cm/hr)
10	35	3	0.03	1.85	4.212
20	35	3	0.07	1.70	2.520
30	32	3	0.07	1.70	2.520
50	34	3	0.03	1.60	0.803
70	31	3	0.03	1.60	0.005
90	32	3	0.05	1.90	0.005
110	32	3	0.05	1.80	0.005
150	30	3	0.05	1.80	0.001

(a)

Sensor Location Below Land Surface (cm)	$\theta_S(\%)$	$\theta_R(\%)$	Φ (cm^{-1})	n(-)	K_S(cm/hr)
10	38	3	0.02	1.35	0.0100
20	34	3	0.03	1.35	0.0100
30	31	3	0.03	1.35	0.0100
50	31	3	0.07	1.90	0.0100
70	31	3	0.2	2.20	0.0100
90	31	3	0.2	2.20	0.0004
110	33	3	0.2	2.20	0.0004
150	35	3	0.2	2.10	0.0012

(b)

Table 2. Soil parameters for study locations in (a) Grassland and (b) Forested area.

Once the soil parameterization is complete root water uptake from each section can be calculated. For any given soil layer in the vertical soil column (**Figure 8**), above the observed water table, observed water content and **Equation 11** can be used to calculate the hydraulic head. For soil layers below the water table hydraulic head is same as the depth of soil layer below the water table due to assumption of hydrostatic pressure. Similarly using **Equation 12** hydraulic conductivity can be calculated. Hence, at any instant in time hydraulic head in each of the eight soil layers can be calculated. To determine total head, gravity head, which is the height of the soil layer above a common datum, has to be added to the hydraulic head.

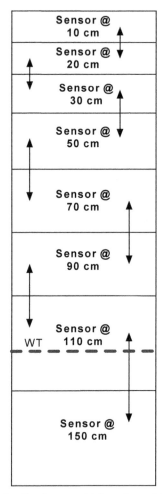

Fig. 8. Schematics of the vertical soil column with location of the soil moisture sensors and water table.

To quantify flow across each soil layer, Darcy's Law (**Equation 7**) is used. Average head values between two consecutive time steps are used to determine the head difference. Also, flow across different soil layers is assumed to be occurring between the midpoints of one layer to another, hence, to determine the head gradient ($\Delta h/l$) the distance between the midpoints of each soil layer is used. The last component needed to solve Darcy's Law is the value of hydraulic conductivity. For flow occurring between layers of different hydraulic conductivities equivalent hydraulic conductivity is calculated by taking harmonic means of the hydraulic conductivities of both the layers (Freeze and Cherry 1979). Hence for each time step harmonically averaged hydraulic conductivity values (**Equation 13**) were used to calculate the flow across soil layers.

$$K_{eq} = \frac{2K_1K_2}{K_1 + K_2} \tag{13a}$$

where K_1 [LT^{-1}]and K_2 [LT^{-1}]are the two hydraulic conductivity values for any two adjacent soil layers and K_{eq} [LT^{-1}]is the equivalent hydraulic conductivity for flow occurring between those two layers.

Figure 9 shows a typical flow layer with inflow and outflow marked. Now using **simple mass balance** changes in water content at two consecutive time steps can be attributed to net inflow minus the root water uptake (assuming no other sink is present). Equation 6.9 can hence be used to determine root water uptake from any given soil layer

$$RWU = (\theta^t - \theta^{t+1}) - (q_{out} - q_{in}) \tag{13b}$$

Using the described methodology one can determine the root water uptake from each soil layer at both study locations (site A and site B).Time step for calculation of the root water uptake was set as four hours and the root water uptake values obtained were summed up to get a daily value for each soil layer.

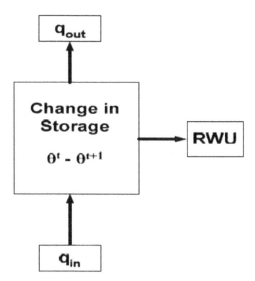

Fig. 9. Schematics of a section of vertical soil column showing fluxes and change in storage.

Using the above methodology root water uptake was calculated from each section of roots for tree and grass land cover from January to December 2003 at a daily time step. **Figure 10** (a and b) shows the variation of root water uptake for a representative period from May 1st to May 15th 2003, This particular period was selected as the conditions were dry and their was no rainfall. Graphs in **Figure 10** (a and b) show the root water uptake variation from section corresponding to each section. Also plotted on the graphs is the normalized water content, which also gives an indication, of water lost from the section.

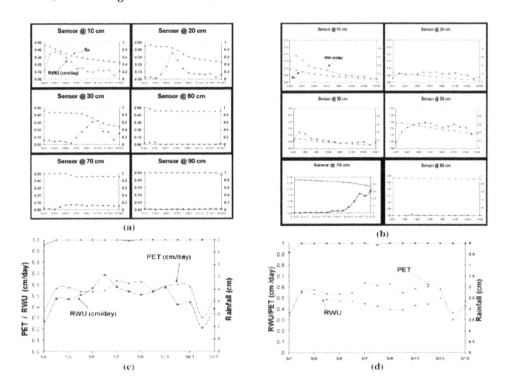

Fig. 10. Root water uptake from sections of soil corresponding to each sensor on the soil moisture instrument for (a, c) Grass land and (b, d) Forest land cover

Figure 10(a) shows the root water uptake from grassed site while panel of graphs in **Figure 10(b)** plots RWU from the forested area. From **Figure 10** (a and b) it can be seen that in both the cases of grass and forest the root water uptake varies with water content and as the top layers starts to get dry, the water uptake from the lower layer increases so as to keep the root water uptake constant clearly indicating that the compensation do take place and hence the models need to account for it. Another important point to note is that in **Figure 10(a)** root water uptake from top three sensors is accounts for the almost all the water uptake while in **Figure 10(b)** the contribution from fourth and fifth sensor is also significant. Also, as will be shown later, in case of forested land cover, root water uptake is observed from the sections that are even deeper than 70 cm below land surface. This is expected owing to the differences in the root system of both land cover types. While grasses have shallow roots, forest trees tend to put their roots deeper into the soil to meet their high water consumptive use.

Figure 10(c and d) show the values of PET plotted along with the observed values of root water uptake. On comparing the grass versus forested graphs it is evident while the grass is still evapotranspiring at values close to PET root water uptake from forested land covers is occurring at less than potential. This behavior can be explained by the fact that water content in the grassed region (as shown by the normalized water content graph, Se) is greater than that of the forest and even though the 70 cm sensor shows significant contribution the uptake is still not sufficient to meet the potential demand.

Figure 11 shows an interesting scenario when a rainfall event occurs right after a long dry stretch that caused the upper soil layers to dry out. **Figure 11(a)** shows the root water uptake profile on 5/18/2003 for forested land cover with maximum water being taken from section of soil profile corresponding to 70 cm below the land surface. A rainfall event of 1inch took place on 5/19/2003. As can be clearly seen in **Figure 11(b)** the maximum water uptake shifts right back up to 10 cm below the land surface, clearly showing that the ambient water content directly and quickly affects the root water uptake distribution. **Figure 11(c)** which shows the snapshot on 5/20/2003 a day after the rainfall where the root water uptake starts redistributing and shifting toward deeper wetter layers. In fact this behavior was observed for all the data analyzed for the period of record for both the grass and forested land covers. With roots taking water from deeper wetter layers and as soon as the shallower layer becomes wet the uptakes shifts to the top layers. **Figure 12** (a and b) show a long duration of record spanning 2 months (starting October to end November), with the whiter shade indicating higher root water uptake. From both the figures it is evident that water uptake significantly shifts in lieu of drier soil layers especially in case of forest land cover (**Figure 12(b)**), while in case of grass uptake is primarily concentrated in the top layers.

As a quick summary the results indicate that

a.   Assuming RWU as directly proportional to root density may not be a good approximation.
b.   Plants adjust to seek out water over the root zone
c.   In case of wet conditions preferential RWU from upper soil horizons may take place
d.   In case of low ET demands the distribution on ET was found to be occurring as per the root distribution, assuming an exponential root distribution

Hence, traditionally used models are not adequate, to model this behavior. Changes in regard to the modeling techniques as well as conceptualizations, hence, need to occur. Plant physiology is one area that needs to be looked into to see what plant properties affect the water uptake and how can they be modeled mathematically. The next section discusses a modeling framework based on plant root characteristics which can be employed to model the aforesaid observations.

5.3 Incorporation of plant physiology in modeling root water uptake

Any framework to model root water uptake dynamically, will have to explicitly account for all the four points listed above. The dynamic model should be able to adjust the uptake pattern based on root density as well as available water across the root zone. The model should use physically based parameters so as to remove empiricism from the formulation of the equations. For a given distribution of water content along the root zone (observed or modeled) knowledge of root distribution as well as hydraulic characteristics of roots is hence essential to develop a physically based root water uptake model. The following two sections will describe how root distributions can be modeled as well as how do roots need to be characterized to model uptake from root's perspective.

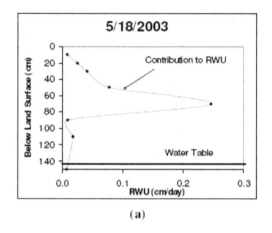

(a)

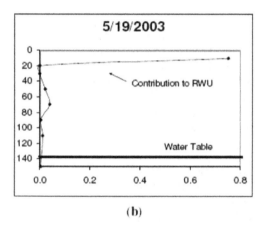

(b)

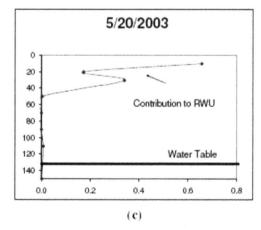

(c)

Fig. 11. Root water uptake variation due to a one inch rainfall even on 5/19/2003.

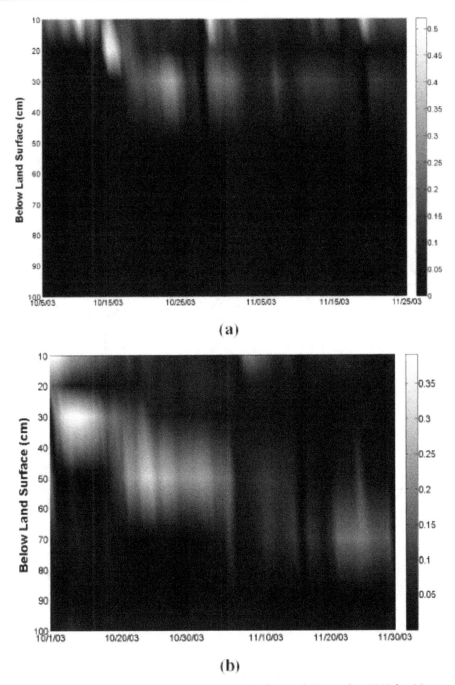

Fig. 12. Daily root water uptake variation for two October and November 2003 for (a) grass land cover and (b) forested land cover.

5.3.1 Root distribution

Schenk and Jackson (2002) expanded an earlier work of Jackson et al. (1996) to develop a global root database having 475 observed root profiles from different geographic regions of the world. It was found that by varying parameter values the root distribution model given by Gale and Grigal (1987) can be used with sufficient accuracy to describe the observed root distributions. **Equation 14** describes the root distribution model.

$$Y = 1 - \gamma^d \tag{14}$$

where Y is the cumulative fraction of roots from the surface to depth d, and γ is a numerical index of rooting distribution which depends on vegetation type. **Figure 13** shows the observed distribution (shown by data points) versus the fitted distribution using **Equation 14** for different vegetation types. The figure clearly indicates the goodness of fit of the above model. Hence, for a given type of vegetation a suitable γ can be used to describe the root distribution.

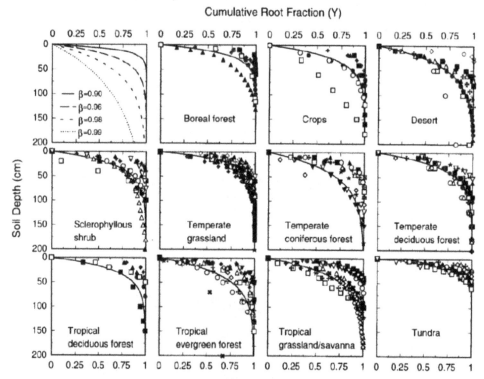

Fig. 13. Observed and Fitted Root Distribution for different type of land covers. [Adapted from Jackson et al. 1996]

5.3.2 Hydraulic characterization of roots

Hydraulically, soil and xylem are similar as they both show a decrease in hydraulic conductivity with reduction in soil moisture (increase in soil suction). For xylem the relationship between hydraulic conductivity and soil suction pressure is called 'vulnerability curve' (Sperry et al. 2003) (see **Figure 14**). The curves are drawn as a percentage loss in conductivity rather than absolute value of conductivity due to the ease of determination of former. Tyree et al (1994) and Hacke et al (2000) have described methods for determination of vulnerability curves for different types of vegetation.

Commonly, the stems and/or root segments are spun to generate negative xylem pressure (as a result of centrifugal force) which results in loss of hydraulic conductivity due to air seeding into the xylem vessels (Pammenter and Willigen 1998). This loss of hydraulic conductivity is plotted against the xylem pressure to get the desired vulnerability curve.

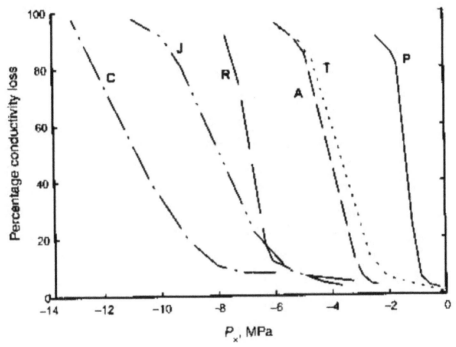

Fig. 14. Vulnerability curves for various species. [Adapted from Tyree, 1999]

For different plant species the vulnerability curve follows an S-Shape function, see **Figure 14** (Tyree 1999). In **Figure 14**, y-axis is percentage loss of hydraulic conductivity induced by the xylem pressure potential Px, shown on the x-axis. C= Ceanothus megacarpus, J = Juniperus virginiana, R = Rhizphora mangel, A = Acer saccharum, T= Thuja occidentalis, P = Populus deltoids.

Pammenter and Willigen (1998) derived an equation to model the vulnerability curve by parametrizing the equation for different plant species. **Equation 15** describes the model mathematically.

$$PLC = \frac{100}{1 + e^{a.(P - P_{50\,PLC})}} \tag{15}$$

where PLC denotes the percentage loss of conductivity P_{50PLC} denotes the negative pressure causing 50% loss in the hydraulic conductivity of xylems, P represents the negative pressure and a is a plant based parameter. **Figure 15** shows the model plotted against the data points for different plants. Oliveras et al. (2003) and references cited therein have parameterize the model for different type of pine and oak trees and found the model to be successful in modeling the vulnerability characteristics of xylem.

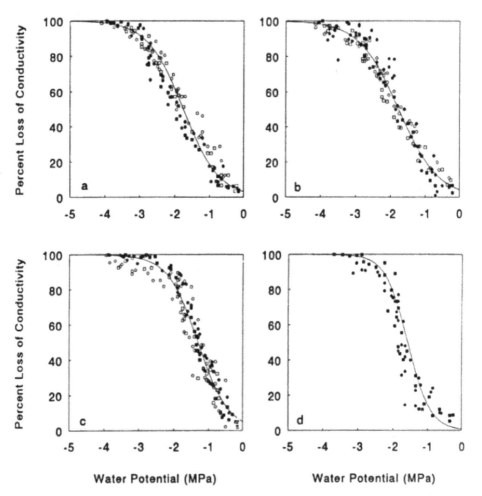

Fig. 15. Observed values and fitted vulnerability curve for roots and stem sections of different Eucylaptus trees. [Adapted from Pammenter and Willigen, 1998].

The knowledge of hydraulic conductivity loss can be used analogous to the water stress response function a (**Equation 9**) by scaling PLC from 0 to 1 and converting the suction pressure to water head. The advantage of using vulnerability curves instead of Feddes or van Genuchten model is that vulnerability curves are based on xylem hydraulics and hence can be physically characterized for each plant species.

5.3.3 Development of a physically based root water uptake model

The current model development is based on model conceptualization proposed by Jarvis (1989) however the parameters for the current model are physically defined and include plant physiological characteristics.

For a given land cover type **Equation 14** and **15** can be parameterize to determine the root fraction for any given segment in root zone and percentage loss of conductivity for a given soil suction pressure. For consistency of representation percentage loss of conductivity will be hence forth represented by α (scaled between 0 and 1 similar to **Equation 9**) and will be called stress index.

For any section of root zone, for example i^{th} section, root fraction can be written as R_i and stress index, determined from vulnerability curve and ambient soil moisture condition, can be written as a_i. Average stress level $\bar{\alpha}$ over the root zone can be defined as the

$$\bar{\alpha} = \sum_{i=1}^{n} R_i \alpha_i \tag{16}$$

where n represents the number of soil layers and the other symbols are as previously defined. Thus, as can be seen from **Equation 16** the average stress level $\bar{\alpha}$ combines the effect of both the root distribution and the available water content (via vulnerability curve).

As shown in **Figure 12(b)** if there is available moisture in the root zone, plant can transpire at potential by increasing the uptake from the lower wetter section of the roots. In terms of modeling it can be conceptualized that above a certain critical average stress level ($\bar{\alpha}_C$) plants can transpire at potential and below $\bar{\alpha}_C$ the value of total evapotranspiration decreases. The decrease in the ET value can be modeled linearly as shown by Li et al (2001). The graph of average stress level versus ET (expresses as a ratio with potential ET rate) can hence be plotted as shown in **Figure 16**. In **Figure 16**, ET_a is the actual ET out of the soil column while ET_p is the potential value of ET. **Figure 16** can be used to determine the value of actual ET for any given average stress level.

Once the actual ET value is known, the contribution from individual sections can be modeled depending on the weighted stress index using the relationship defined by

$$S_i = \left(\frac{E_a}{\Delta Z_i} \right) \left(\frac{R_i \alpha_i}{\bar{\alpha}} \right) \tag{17}$$

where S_i defined as the water uptake from the ith *section,* ΔZ_i is the depth of ith section and other symbols are as previously defined

Jarvis (1989) used empirical values to simulate the behavior of the above function and **Figure 17** shows the result of root water uptake obtained from his simulation. The values next to each curve in **Figure 17** represent the day after the start of simulation and actual ET rate as expressed in mm/day. On comparison with **Figure 12**, the model successfully reproduced the shift in root water uptake pattern with the uptake being close to potential value (ET_P = 5.0 mm/d) for about a month from the start of simulation. The decline in ET rate occurred long after the start of the simulation in accordance with the observed values. The model was successful not only in simulating peak but also in the observed magnitude of the root water uptake.

From the above analysis it can be concluded that the root water uptake is just not directly proportional to the distribution of the roots but also depends on the ambient water content. Under dry conditions roots can easily take water from deeper wetter soil layers.

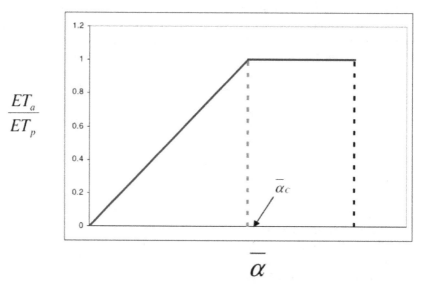

Fig. 16. Variation of ratio of actual to potential ET with location of the critical stress level.

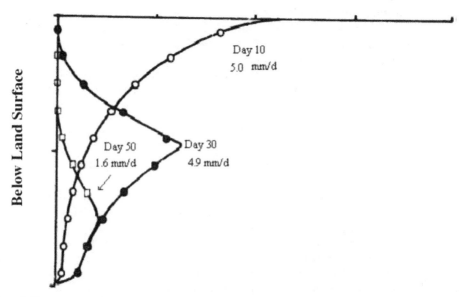

Fig. 17. Variation in the vertical distribution of root water uptake, at different times. [Adapted from Jarvis (1989)]

The methodology described here involves initial laboratory analyses to determine the hydraulic characteristics of the plant. However, once a particular plant specie is characterized then the parameters can be use for that specie elsewhere under similar conditions. The approach shows that eco-hydrological framework has great potential for improving predictive hydrological modeling.

6. Conclusion

The chapter described a method of data collection for soil moisture and water table that can be used for estimation of evapotranspiration. Also described in the chapter is the use of vertical soil moisture measurements to compute the root water uptake in the vadose zone and use that uptake to validate a root water uptake model based on plant physiology based root water uptake model. As evaporation takes place primarily from the first few centimeters (under normal conditions) of the soil profile and the biggest component of the ET is the root water uptake. Hence to improve our estimates of ET, which constitutes ~70% of the rainfall, the estimation and modeling of root water uptake needs to be improved. Eco-hydrology provides one such avenue where plant physiology can be incorporated to better represent the water loss. Also, hydrological model incorporating plant physiology can be modified easily in future to be used to predict land-cover changes due to changes in rainfall pattern or other climatic variables.

7. References

Allen RG, Pereira LS, Raes D, Smith M.1998. Crop evapotranspiration—guidelines for computing crop water requirements.FAO Irrigation & Drainage Paper 56. FAO, Rome

Bidlake, W. R., W.M.Woodham, and M.A.Lopez. 1993. Evapotranspiration from areas of native vegetation in Wets-Central Florida: U.S. Geological Survey open file report 93-415, 35p.

Brutsaert, W.1982. Evaporation into the Atmosphere: Theory, History, and Applications. Kluwer Academic Publishers, Boston, MA.

Doorenbos, J., and W.O.Pruitt. 1977. Crop Water Requirements. FAO Irrigation and drainage paper 24. Food and agricultural organization of the United Nations, Rome.

Fares, A. and A.K. Alva. 2000. Evaluating the capacitance probes for optimal irrigation of citrus through soil moisture monitoring in an Entisol profile. Irrigation. Science 19:57–64.

Fayer,M.J. and D.Hillel.1986. Air Encapsulation I - Measurement in a field soil. Soil Science Society of America Journal. 50:568-572.

Feddes,R.A., P.J.Kowalik, and H.Zaradny. 1978. Simulation of field water use and crop yield. New York: John Wiley & Sons.

Freeze,R. and J.Cherry. 1979. Groundwater. Prentice Hall, Old Tappan, NJ.

Hacke.U.G., J.S.Sperry, and J.Pittermann. 2000. Drought Experience and Cavitation Resistance in Six Shrubs from the Great Basin, Utah. Basic Applied Ecology 1:31-41.

Hillel,D. 1998. Environmental soil physics. Academic Press, New York, NY

Jackson, R.B., J.Canadell, J.R.Ehleringer, H.A.Mooney, O.E.Sala, and E.D.Schulze. 1996. A global analysis of root distributions for terrestrial biomes. Oecologia 108:389-411.

Jarvis.N.J. 1989. A Simple Empirical Model of Root Water Uptake. Journal of Hydrology.107:57-72.

Kite, G.W., and P. Droogers. 2000. Comparing evapotranspiration estimates from satellites, hydrological models and field data. Journal of Hydrology 229:3-18.

Knowles, L., Jr. 1996. Estimation of evapotranspiration in the Rainbow Springs and Silver Springs basin in north-central Florida. Water resources investigation report. 96-4024. USGS, Reston, VA.

Li, K.Y., R.De jong, and J.B. Boisvert. 2001. An exponential root-water-uptake model with water stress compensation. Journal of hydrology 252:189-204.

Li,K.Y., R.De Jong, and M.T.Coe. 2006. Root water uptake based upon a new water stress reduction and an asymptotic root distribution function. Earth Interactions 10 (paper 14):1-22.

Mahmood, R. and K.G. Hubbard. 2003. Simulating sensitivity of soil moisture and evapotranspiration under heterogeneous soils and land uses. Journal of Hydrology. 280:72-90.

Meyboom, P. 1967. Ground water studies in the Assiniboine river drainage basin: II. Hydrologic characteristics of phreatophytic vegetation in south-central Saskatchewan. Geological Survey of Canada Bulletin 139, no.64.

Monteith, J. L. 1965. Evaporation and environment. In G.E.Fogg (ed). The state and movement of water in living organisms. Symposium of the Society of Experimental Biology: San Diego, California, Academic Press, New York, p.205-234

Morgan,K.T., L.R.Parsona, T.A. Wheaton, D.J.Pitts and T.A.Oberza. 1999. Field calibration of a capacitance water content probe in fine sand soils. Soil Science Society of America Journal 63: 987-989.

Mo, X., S. Liu, Z. Lin, and W. Zhao. 2004. Simulating temporal and spatial variation of evapotranspiration over the Lushi basin. Journal of Hydrology 285:125-142.

Mualem, Y. 1976. A new model predicting the hydraulic conductivity of unsaturated porous media. Water Resources Research 12(3):513-522.

Nachabe, M., N.Shah, M.Ross, and J.Vomacka. 2005. Evapotranspiration of two vegetation covers in a shallow water table environment. Soil Science Society of America Journal 69:492-499.

Oliveras,I., J.Martinez-Vilalta, T.Jimenez-Ortiz, M.J Lledo, A.Escarre, and J.Pinol. 2003. Hydraulic Properties of Pinus Halepensis, Pinus Pinea, and Tetraclinis Articulata in a Dune Ecosystem of Eastern Spain. Plant Ecology 169:131-141.

Pammenter.N.W. and C.V.Willigen. 1998. A Mathematical and Statistical Analysis of the Curves Illustrating Vulnerability of Xylem to Cavitation. Tree Physiology 18:589-593.

Priestley, C.H.B., and Taylor, R.J. 1972. On the assessment of surface heat flux and evaporation using large-scale parameters. Monthly Weather Review 100(2): 81-92.

Richards.L.A .1931. Capillary conduction of liquids through porous mediums, Journal of Applied Physics, 1(5), 318-333.

Said , A., M.Nachabe, M.Ross, and J.Vomacka. 2005. Methodology for estimating specific yield in shallow water environment using continuous soil moisture data. ASCE Journal of Irrigation and Drainage Engineering 131, no.6:533-538.

Schenk, H.J. and R. B. Jackson. 2002. Rooting Depths, Lateral Root Spreads and Below-Ground/Above-Ground Allometries of Plants in Water-Limited Ecosystems. The Journal of Ecology 90(3):480-494.

Shah,N. 2007. Vadose Zone Processes Affecting Water Table Fluctuations – Conceptualization and Modeling Considerations. PhD Disseration, University of South Florida, Tampa, Fl, 233 pp.

Shah, N., M. Ross, and A. Said. 2007. Vadose Zone Evapotranspiration Distribution Using One-Dimensional Analysis and Conceptualization for Integrated Modeling. Proceedings of ASCE EWRI conference, May 14th –May 19th 2007, Tampa.

Simunek, J., M. Th. van Genuchten and M. Sejna. 2005. The HYDRUS-1D software package for simulating the movement of water, heat, and multiple solutes in variably saturated media, version 3.0, HYDRUS software series 1. Department of Environmental Sciences, University of California Riverside, Riverside, California, USA, 270 pp.

Sperry, J.S., V.Stiller, and U.G.Hacke. 2003 Xylem Hydraulics and the Soil-Plant-Atmosphere Continuum: Oppurtunities and Unresolved Issues. Agronomy Journal 95:1362-1370.

Sumner, D.M. 2001. Evapotranspiration from a cypress and pine forest subjected to natural fires, Volusia County, Florida, 1998-99. Water Resources Investigations Report 01-4245. USGS, Reston, VA.

Sumner, D. 2006.Adequacy of selected evapotranspiration approximations for hydrological simulation. Journal of the American Water Resources Association. 42(3):699- 711.

Thornthwaite, C.W. 1948. An approach toward a rational classification of climate. Geographic Review 38:55-94.

Trout, K., and M.Ross. 2004. Intensive hydrologic data collection in as small watershed in West-Central Florida. Hydrological Science and Technology 21(1-4):187-197.

Tyree, M.T. S.Yang, P.Cruiziat, and, B.Sinclair. 1994. Novel Methods of Measuring Hydraulic Conductivity of Tree Root Systems and Interpretation Using AMAIZED. Plant Physilogy 104:189-199

van Genuchten, M.Th. 1980. A closed-form equation for predicting the hydraulic conductivity of unsaturated soils. Soil Science Society of America Journal 44:892-898.

van Genuchten, M. Th.1987. A numerical model for water and solute movement in and below the root zone. Research report No 121, U.S. Salinity laboratory, USDA, ARS, Riverside, California, 221pp.

Yang, J., B. Li, and S. Liu. 2000. A large weighing lysimeter for evapotranspiration and soil water-groundwater exchange studies. Hydrological Processes 14:1887-1897.

White,W.N. 1932.A method of estimating ground-water supplies based on discharge by plants and evaporation from soil: Results of investigation in Escalante Valley, Utah. Water-Supply Paper 659-A.

8

Estimation of Evapotranspiration Using Soil Water Balance Modelling

Zoubeida Kebaili Bargaoui
Tunis El Manar University
Tunisia

1. Introduction

Assessing evapotranspiration is a key issue for natural vegetation and crop survey. It is a very important step to achieve the soil water budget and for deriving drought awareness indices. It is also a basis for calculating soil-atmosphere Carbon flux. Hence, models of evapotranspiration, as part of land surface models, are assumed as key parts of hydrological and atmospheric general circulation models (Johnson et al., 1993). Under particular climate (represented by energy limiting evapotranspiration rate corresponding to potential evapotranspiration) and soil vegetation complex, evapotranspiration is controlled by soil moisture dynamics. Although radiative balance approaches are worth noting for evapotranspiration evaluation, according to Hofius (2008), the soil water balance seems the best method for determining evapotranspiration from land over limited periods of time. This chapter aims to discuss methods of computing and updating evapotranspiration rates using soil water balance representations.

At large scale, Budyko (1974) proposed calculating annual evapotranspiration from data of meteorological stations using one single parameter w_0 representing a critical soil water storage. Using a statistical description of the sequences of wet and dry days, Eagleson (1978 a) developed an average annual water balance equation in terms of 23 variables including soil, climate and vegetation parameters with the assumption of a homogeneous soil-atmosphere column using Richards (1931) equation. On the other hand, the daily bucket with bottom hole model (BBH) proposed by Kobayashi et al. (2001) was introduced based on Manabe model (1969) involving one single layer bucket but including gravity drainage (leakage) as well as capillary rise. Vrugt et al. (2004) concluded that the daily Bucket model and the 3-D model (MODHMS) based on Richards equation have similar results. Also, Kalma & Boulet (1998) compared simulation results of the rainfall runoff hydrological model VIC which assumes a bucket representation including spatial variability of soil parameters to the one dimensional physically based model SiSPAT (Braud et al. , 1995). Using soil moisture profile data for calibration, they conclude that catchment's scale wetness index for very dry and very wet periods are misrepresented by SiSPAT while captured by VIC. Analyzing VIC parameter identifiability using streamflow data, DeMaria et al. (2007) concluded that soil parameters sensitivity was more strongly dictated by climatic gradients than by changes in soil properties especially for dry environments. Also, studying the measurements of soil moisture of sandy soils under semi-arid conditions, Ceballos et al. (2002) outlined the dependence of soil moisture time series on intra annual rainfall variability. Kobayachi et al. (2001) adjusted soil humidity profiles measurements for model calibration while Vrugt et al. (2004) suggested that effective soil hydraulic properties are poorly identifiable using drainage discharge data.

The aim of the chapter is to provide a review of evapotranspiration soil water balance models. A large variety of models is available. It is worth noting that they do differ with respect to their structure involving empirical as well as conceptual and physically based models. Also, they generally refer to soil properties as important drivers. Thus, the chapter will first focus on the description of the water balance equation for a column of soil-atmosphere (one dimensional vertical equation) (section 2). Also, the unsaturated hydrodynamic properties of soils as well as some analytical solutions of the water balance equation are reviewed in section 2. In section 3, key parameterizations generally adopted to compute actual evapotranspiration will be reported. Hence, several soil water balance models developed for large spatial and time scales assuming the piecewise linear form are outlined. In section 4, it is focused on rainfall-runoff models running on smaller space scales with emphasizing on their evapotranspiration components and on calibration methods. Three case studies are also presented and discussed in section 4. Finally, the conclusions are drawn in section 5.

2. The one dimensional vertical soil water balance equation

As pointed out by Rodriguez-Iturbe (2000) the soil moisture balance equation (mass conservation equation) is "likely to be the fundamental equation in hydrology". Considering large spatial scales, Sutcliffe (2004) might agree with this assumption. In section 2.1 we first focus on the presentation of the equation relating relative soil moisture content to the water balance components: infiltration into the soil, evapotranspiration and leakage. Then water loss through vegetation is addressed. Finally, infiltration models are discussed in section 2.2.

2.1 Water balance

For a control volume composed by a vertical soil column, the land surface, and the corresponding atmospheric column, and under solar radiation and precipitation as forcing variables, this equation relates relative soil moisture content s to infiltration into the soil $I(s,t)$, evapotranspiration $E(s,t)$ and leakage $L(s,t)$.

$$nZ_a \, \delta s/\delta t = I(s,t) - E(s,t) - L(s,t) \qquad (1a)$$

Where t is time, n is soil effective porosity (the ratio of volume of voids to the total soil matrix volume); and Z_a is the active depth of soil.

Soil moisture exchanges as well as surface heat exchanges depend on physical soil properties and vegetation (through albedo, soil emissivity, canopy conductance) as well as atmosphere properties (turbulent temperature and water vapour transfer coefficients, aerodynamic conductance in presence of vegetation) and weather conditions (solar radiation, air temperature, air humidity, cloud cover, wind speed). Soil moisture measurements require sampling soil moisture content by digging or soil augering and determining soil moisture by drying samples in ovens and measuring weight losses; also, in situ use of tensiometry, neutron scattering, gamma ray attenuation, soil electrical conductivity analysis, are of common practice (Gardner et al. (2001) ; Sutcliffe, 2004; Jeffrey et al. (2004)).

The basis of soil water movement has been experimentally proposed by Darcy in 1856 and expresses the average flow velocity in a porous media in steady-state flow conditions of groundwater. Darcy introduced the notion of hydraulic conductivity. Boussinesq in 1904 introduced the notion of specific yield so as to represent the drainage from the unsaturated zone to the flow in the water table. The specific yield is the flux per unit area draining for a unit fall in water table height. Richards (1931) proposed a theory of water movement in the unsaturated homogeneous bare soil represented by a semi infinite homogeneous column:

$$\delta\theta/\delta t = \delta/\delta z \left[K \, \delta\psi/\delta z - K(\theta) \right] \tag{1b}$$

Where t is time; θ is volumetric water content (which is the ratio between soil moisture volume and the total soil matrix volume cm^3cm^{-3}); z is the vertical coordinate ($z>0$ downward from surface); K is hydraulic conductivity (cms^{-1}); ψ is the soil water matrix potential. Both K and ψ are function of the volumetric water content. Richards equation assumes that the effect of air on water flow is negligible. If accounting for the slope surface, it comes:

$$\delta\theta/\delta t = \delta/\delta z \left[K \, \delta\psi/\delta\theta \, \delta\theta/\delta z\right] - \delta \, K/\delta\theta \, \delta\theta/\delta z] \cos\beta \tag{2}$$

Where β is surface slope angle and cos is the cosinus function. We notice that the term [K $\delta\psi/\delta z$ - K(θ)] represents the vertical moisture flux. In particular, as reported by Youngs (1988) the soil-water diffusivity parameter D has been proposed by Childs and Collis-George (1950) as key soil-water property controlling the water movement.

$$D(\theta) = K(\theta) \, \delta\psi/\delta\theta \tag{3}$$

Thus, the Richards equation is often written as following:

$$\delta\theta/\delta t = \delta/\delta z \left[D(\theta) \, \delta\theta/\delta z \right] - \delta \, K(\theta)/\delta z] \tag{4}$$

Eq. (4) is generally completed by source and sink terms to take into account the occurrence of precipitation infiltrating into the soil $I_{nf}(\theta,z_0)$ where z_0 is the vertical coordinate at the surface and vegetation uptake of soil moisture $g_r(\theta,z)$,. Vegetation uptake (transpiration) depends on vegetation characteristics (species, roots, leaf area, and transfer coefficients) and on the potential rate of evapotranspiration E_0 which characterizes the climate. Consequently, Eq. (4) becomes:

$$\delta\theta/\delta t = \delta/\delta z \left[D(\theta)\delta\theta/\delta z - K(\theta) \right] - g_r(\theta,z) + I_{nf}(\theta,z_0) \tag{5}$$

Youngs (1988) noticed that near the soil surface where temperature gradients are important Richards equation may be inadequate. We find in Raats (2001) an important review of evapotranspiration models and analytical and numerical solutions of Richards equation. However, it should be noticed that after Feddes et al. (2001) "in case of catchments with complex sloping terrain and groundwater tables, a vertical domain model has to be coupled with either a process or a statistically based scheme that incorporates lateral water transfer". So, a key task in the soil water balance model evaluation is the estimation of $I_{nf}(\theta,z_0)$ and $g_r(\theta,z)$. Both depend on the distribution of soil moisture. We focus here on vegetation uptake (or transpiration) $g_r(\theta,z)$ which is regulated by stomata and is driven by atmospheric demand. Based on an Ohm's law analogy which was primary proposed by Honert in 1948 as outlined by Eagleson (1978 b), the conceptual model of local transpiration uptake u(z,t)= $g_r(\theta,z)$ as volume of water per area per time is expressed as (Guswa, 2005)

$$u(z,t) = \Delta z \, (\psi(z,t) - \psi_p) \, / [\, R_1(\theta \, (z,t)) + R_2] \tag{6}$$

ψ soil moisture potential (bars), ψ_p leaf moisture potential (bars); R_1 (s cm^{-1}) a resistance to moisture flow in soil; it depends on soil and root characteristics and is function of the volumetric water content; R_2 (s cm^{-1}) is vegetation resistance to moisture flow; Δz is soil depth. It is worth noting that $\psi_p > \psi*$ where $\psi*$ is the wilting point potential; In Ceballos et al. (2002) the wilting point is taken as the soil-moisture content at a soil-water potential of - 1500 kPa.

Estimations of air and canopy resistances R_1 and R_2 often use semi-empirical models based on meteorological data such as wind speed as explanatory variables (Monteith (1965); Villalobos et al., 2000). Jackson et al. (2000) pointed out the role of the Hydraulic Lift process which is the movement of water through roots from wetter, deeper soil layers into drier, shallower layers along a gradient in ψ. On the basis of such redistribution at depth, Guswa (2005) introduced a parameter to represent the minimum fraction of roots that must be wetted to the field capacity in order to meet the potential rate of transpiration. The field capacity is defined as the saturation for which gravity drainage becomes negligible relative to potential transpiration (Guswa, 2005). The potential matrix at field capacity is assumed equal to 330 hPa (330 cm) (Nachabe, 1998). The resulting $u(z,t)$ function is strongly non linear versus the average root moisture with a relative insensitivity to changes in moisture when moisture is high and sensitivity to changes in moisture when the moisture is near the wilting point conditions. We also emphasize the Perrochet model (Perrochet, 1987) which links transpiration to potential evapotranspiration E_0 through:

$$g_r(\theta,z,t) = \alpha(\theta) r(z) \, E_0(t) \tag{7}$$

Where $r(z)$ (cm^{-1}) is a root density function which depends both on vegetation type and climatic conditions, $\alpha(\theta)$ is the root efficiency function. Both $r(z)$ and $\alpha(\theta)$ represent macroscopic properties of the root soil system; they depend on layer thickness and root distribution . Lai and Katul (2000) and Laio (2006) reported some models assigned to $r(z)$ which are linear or non linear. As out pointed by Laio (2006), models generally assume that vegetation uptake at a certain depth depends only on the local soil moisture. It is noticeable that in Feddes et al. (2001), a decrease of uptake is assumed when the soil moisture exceeds a certain limit and transpiration ceases for soil moisture values above a limit related to oxygen deficiency.

2.2 Review of models for hydrodynamic properties of soils
Many functional forms are proposed to describe soil properties evolution as a function of the volumetric water content (Clapp et al. , 1978). They are called retention curves or pedo transfer functions. We first present the main functional forms adopted to describe hydraulic parameters (section 2.2.1). Then, we report some solutions of Richards equation (section 2.2.2).

2.2.1 Functional forms of soil properties
According to Raats (2001), four classes of models are distinguishable for representing soil hydraulic parameters. Among them the linear form with D as constant and K linear with θ and the function Delta type as proposed by Green Ampt $D = \frac{1}{2} s^2 \, (\theta_1 - \theta_0)^{-1} \, \delta(\theta_1 - \theta_0)$ where s is the degree of saturation (which is the ratio between soil moisture volume and voids

volume; s=1 in case of saturation) and θ_1; θ_0 parameters. Also power law functions for $\psi(\theta)$ and $K(\theta)$ are proposed by Brooks and Corey (1964) on the basis of experimental observations while Gardner (1958) assumes exponential functions. The power type model proposed by Brooks & Corey (1964) are the most often adopted forms in rainfall-runoff transformation models. The Brooks and Corey model for K and ψ is written as:

$$K(s) = K(1) \, s^{c'} \; ; \; \psi(s) = \psi(1) \, s^{-1/m} \tag{8}$$

where m is a pore size index and c' a pore disconnectedness index (Eagleson 1978 a,b); After Eagleson (1978a, b), c' is linked to m with $c'=(2+3m)/m$. In Eq. (8), $K(1)$ is hydraulic conductivity at saturation (for s=1); $\psi(1)$ is the bubbling pressure head which represents matrix potential at saturation. During dewatering of a sample, it corresponds to the suction at which gas is first drawn from the sample; As a result, Brooks and Corey (BC) model for diffusivity is derived as:

$$D(\theta) = s^d \, \psi(1) \; K(1) \, / (nm) \tag{9}$$

where n is effective soil porosity; and $d=(c'-1-(1/m))$. Let's consider the intrinsic permeability k which is a soil property. (K and k are related by $K = k \, \rho_w / \mu$ where μ dynamic viscosity of water; ρ_w specific weight of pore water). After Eagleson (1978 a, b), three parameters involved in pedo transfer functions may be considered as independent parameters: n, c' and k(1) where k(1) is intrinsic permeability at saturation.

On the other hand, Gardner (1958) model assumed the exponential form for the hydraulic conductivity parameter (Eq. 10):

$$K(\psi) = K_S \, e^{-a' \, \psi} \tag{10}$$

Where K_S saturated hydraulic conductivity at soil surface; a' pore size distribution parameter. Also, in Gardner (1958) model, the degree of saturation and the soil moisture potential are linked according to Eq. (11). The power function introduces a parameter l which is a factor linked to soil matrix tortuosity ($l = 0.5$ is recommended for different types of soils);.

$$s(\psi) = [e^{-0.5 \, a' \, \psi} \, (1 + 0.5 \, a' \, \psi)]^{2/(l+2)} \tag{11}$$

Van Genutchen model (1980) is another kind of power law model but it is highly non linear

$$K(\psi) = K_S \, s^{\lambda+1} \, [\, 1 - (1 - s^{(\lambda+1)/\lambda})^{\lambda/(\lambda+1)}\,]^2 \tag{12}$$

$$s(\psi) = [1 + (\psi(1)/\psi))^{-(\lambda+1)}\,]^{-\lambda/(\lambda+1)} \text{ for } \psi \leq \psi(1);$$

$$s = 1 \qquad\qquad \text{ for } \psi < \psi(1) \tag{13}$$

In Eq. (12) and (13) λ is a parameter to be calibrated. Calibration is generally performed on the basis of the comparison of computed and observed retention curves.

In order to determine K_S one way is to adopt Cosby et al. (1984) model (Eq. 14).

$$\text{Log}(K_S) = -0.6 + (0.0126 \, S_\% - 0.0064 \, C_\%) \tag{14}$$

Where $S_\%$ and $C_\%$ stand for soil percents of sand and clay. Also, we may find tabulated values of K_S (in m/day) according to soil texture and structure properties in FAO (1980). On the other hand, soil field capacity S_{FC} plays a key role in many soil water budget models. In Ceballos et al. (2002) the field capacity was considered as "the content in humidity corresponding to the inflection point of the retention curve before it reached a trend parallel to the soil water potential axis". In Guswa (2005), it is defined as the saturation for which gravity drainage becomes negligible relative to potential transpiration. As pointed out by Liao (2006) who agreed with Nachabe (1998), there is an "intrinsic subjectivity in the definition of field capacity". Nevertheless, many semi-empirical models are offered in the literature for S_{FC} estimation as a function of soil properties (Nachabe, 1988). In Cosby (1984), S_{FC} expressed as a degree of saturation is assumed s:

$$S_{FC} = 50.1 + (-0.142\ S_\% - 0.037\ C_\%) \tag{15}$$

On the other hand, according to Cosby (1984) and Saxton et al. (1986) S_{FC} may be derived as:

$$S_{FC} = (20/A')^{1/B'} \tag{16}$$

where
$A' = 100*\exp(a_1 + a_2 C_\% + a_3 S_\%^2 + a_4 S_\%^2 C_\%)$; $B' = a_5 + a_6 C_\%^2 + a_7 S_\%^2 + a_8 S_\%^2 C_\%$; $a_1 = -4{,}396$; $a_2 = -0{,}0715$; $a_3 = -0{,}000488$; $a_4 = -0{,}00004285$; $a_5 = -0{,}00222$; $a_6 = -0{,}00222$; $a_7 = -0{,}00003484$; $a_8 = -0{,}00003484$
Recently, this model was adopted by Zhan et al. (2008) to estimate actual evapotranspiration in eastern China using soil texture information. Also, soil characteristics such as S_{FC} may be obtained from Rawls & Brakensiek (1989) according to soil classification (Soil Survey Division Staff, 1998). Nasta et al. (2009) proposed a method taking advantage of the similarity between shapes of the particle-size distribution and the soil water retention function and adopted a log-Normal Probability Density Function to represent the matrix pressure head function retention curve.

2.2.2 Review of analytical solutions of the movement equation
Two well-known solutions of Richards equation are reported here (Green &Ampt model (1911), Philip model (1957)) as well as a more recent solution proposed by Zhao and Liu (1995). These solutions are widely adopted in rainfall-runoff models to derive infiltration.
In the Green &Ampt method (1911), it is assumed that infiltration capacity f from a ponded surface is:

$$f = K_{av} (1 + \Delta\psi\ \Delta\theta\ F^{-1}) \tag{17}$$

K_{av} average saturated hydraulic conductivity ; $\Delta\psi$ difference in average matrix potential before and after wetting; $\Delta\theta$ difference in average soil water content before and after wetting; F the cumulative infiltration for a rainfall event (with $f = dF/dt$).
In the Philip (1957) solution, it is assumed that the gravity term is negligible so that $\delta K(\theta)/\delta z] \approx 0$. A time series development considers the soil water profile of the form:

$$z(\theta,t) = f_1 (\theta)\ t^{1/2} + f_2 (\theta)\ t + f_3 (\theta)\ t^{3/2} + \ldots \tag{18}$$

Where $f_1, f_2, \ldots$ are functions of θ. Hence, the cumulative infiltration $\Omega_f (t)$ is:

$$\Omega_f (t) = S\ t^{1/2} + (A_2 + K_S)\ t + A_3\ t^{3/2} + \ldots \tag{19}$$

Where S soil sorptivity, K_S is saturated hydraulic conductivity of the soil and $A_1, A_2, \ldots$ are parameters. Philip suggested adopting a truncation that results in:

$$\Omega_f(t) = S\, t^{1/2} + K_S / n'\, t \qquad (20)$$

Where n' is a factor $0.3 < n' < 0.7$. It is worth noting that the soil sorptivity S depends on initial water content. So it has to be adjusted for each rainfall event. This is usually performed by comparing observed and simulated cumulative infiltration. For further discussion of Philip model, the reader may profitably refer to Youngs (1988).

Another model of infiltration is worth noting. It is the model of Zhao and Liu (1995) which introduced the fraction of area under the infiltration capacity:

$$i(t) = i_{max}\,[1 - (1-A(t))^{1/b''}] \qquad (21)$$

Where i(t) is infiltration capacity at time t. Its maximum value is i_{max}. A(t) is the fraction of area for which the infiltration capacity is less than i(t) and b'' is the infiltration shape parameter. As out pointed by DeMaria et al. (2007), the parameter b'' plays a key role. Effectively, an increase in b'' results in a decrease in infiltration.

3. Review of various parameterizations of actual evapotranspiration

Many early works on radiative balance combination methods for estimating latent heat using Penman – Monteith method (Monteith, 1965) were coupled with empirical models for representing the conductance of the soil-plant system (the conductance is the inverse function of the resistance). Based on observational evidence, these works have assumed a linear piecewise relation between volumetric soil moisture and actual evapotranspiration. Thus, several water balance models have been developed for large spatial and time scales assuming this piecewise linear form beginning from the work of Budyko in 1956 as pointed out by Manabe (1969)), Budyko (1974), Eagleson (1978 a, b), Entekhabi & Eagleson (1989) and Milly (1993). In fact, soil water models for computing actual evapotranspiration differ according to the time and space scales and the number of soil layers adopted as well as the degree of schematization of the water and energy balances. Moreover, specific canopy interception schemes, pedo transfer sub-models and runoff sub-models often distinguish between actual evapotranspiration schemes. Also, models differ by the consideration of mixed bare soil and vegetation surface conditions or by differencing between vegetation and soil cover. In the former, there is a separation between bare soil evapotranspiration and vegetation transpiration as distinct terms in the computation of evapotranspiration. In the following, we first present a brief review of land surface models which fully couple energy and mass transfers (section 3.1). Then, we make a general presentation of soil water balance models based on the actualisation of soil water storage in the upper soil zone assuming homogeneous soil (section 3.2).Further, it is focused on the estimation of long term actual evapotranspiration using approximation of the solution of the water balance model (section 3.3). In section 3.4, large scale soil water balance models (bucket schematization) are outlined with much more details. Finally a discussion is performed in section 3.5.

3.1 Review of land surface models

In Soil-Vegetation-Atmosphere-Transfer (SVAT) models or land surface models, energy and mass transfers are fully coupled solving both the energy balance (net radiation equation, soil heat fluxes, sensible heat fluxes, and latent heat fluxes) in addition to water movement equations. Usually this is achieved using small time scales (as for example one hour time increment). The specificity of SVAT models is to describe properly the role of vegetation in the evolution of water and energy budgets. This is achieved by assigning land type and soil information to each model grid square and by considering the physiology of plant uptake. Many SVAT models have been developed in the last 25 years. We may find in Dickinson and al. (1986) perhaps one of the first comprehensive SVAT models which was addressed to be used for General circulation modelling and climate modelling. It was called BATS (Biosphere-Atmosphere Transfer Scheme). It was able to compute surface temperature in response to solar radiation, water budget terms (soil moisture, evapotranspiration and, runoff), plant water budget (interception and transpiration) and foliage temperature. ISBA model (Noilhan et Mahfouf, 1996) was further developed in France and belongs to "simple models with mono layer energy balance combined with a bulk soil description" (after Olioso et al. (2002)). An example of using ISBA scheme is presented in Olioso et al. (2002). The following variables are considered: surface temperature, mean surface temperature, soil volumetric moisture at the ground surface, total soil moisture, canopy interception reservoir. The soil volumetric moisture at the ground surface is adopted to compute the soil evaporation while the total soil moisture is used to compute transpiration. The total latent heat is assumed as a weighted average between soil evaporation and transpiration using a weight coefficient depending on the degree of canopy cover. Canopy albedo and emissivity, vegetation Leaf area index LAI, stomatal resistance, turbulent heat and transfer coefficients are parameters of the energy balance equations. It is worth noting that soil parameters in temperature and moisture are computed using soil classification databases. Without loss of generality we briefly present the two layers water movement model adopted by Montaldo et al. (2001)

$$\delta\theta_g /\delta t = C_1/ (\rho_w d_1) [P_g - E_g] - C_2/\tau [\theta_g - \theta_{geq}] \qquad 0 \le \theta_g \le \theta_s \qquad (22)$$

$$\delta\theta_2 /\delta t = C_1/ (\rho_w d_2) [P_g - E_g - E_{tr} - q_2] \qquad 0 \le \theta_2 \le \theta_s \qquad (23)$$

d_1 and d_2 depth of near surface and root zone soil layers; ρ_w density of the water; θ_g and θ_2 volumetric water contents of near surface and root zone soil layers; θ_{geq} equilibrium surface volumetric soil moisture content ideally describing a reference soil moisture for which gravity balances capillary forces such that no flow crosses the bottom of the near surface zone of depth d_1; P_g precipitation infiltrating into the soil; E_g bare soil evaporation rate at the surface; E_{tr} transpiration rate from the root zone of depth d_2; q_2 rate of drainage out of the bottom of the root zone; It is assumed to be equal to the hydraulic conductivity of the root zone at $\theta = \theta_2$. τ ; C_1 and C_2 are parameters. In this model, the rescaling of the root zone soil moisture θ_2 seems to be highly recommended in order to achieve adequate prediction of θ_g in comparison to observations (Montaldo et al. (2001)). Using an assimilation procedure, Montaldo et al. (2001) achieved overcoming misspecification of K_S of two orders magnitude in the simulation of θ_2.

According to Franks et al. (1997), the calibration of SVAT schemes requires a large number of parameters. Also, field experimentations needed to calibrate these parameters are rather important. Moreover up scaling procedures are to be implemented. Boulet and al. (2000) argued that "detailed SVAT models especially when they exhibit small time and space steps are difficult to use for the investigation of the spatial and temporal variability of land surface fluxes".

3.2 Review of average long term evapotranspiration or "regional" evapotranspiration models

Considering the soil water balance at monthly time scale, Budyko (1974) introduced one single parameter which is a critical soil water storage w_0 corresponding to 1 m homogeneous soil depth. According to Budyko (1974), w_0 is a regional parameter seasonally constant and essentially depending on the climate-vegetation complex. The main assumption is that monthly actual evapotranspiration starts from zero and is a piecewise linear function of the degree of saturation expressed as the ratio w/w_0 where w is the actual soil water storage. Either, for $w \geq w_0$ actual evapotranspiration is assumed at potential value E_0.

Average annual water balance equation is also developed in Eagleson (1978 a) in terms of 23 variables (six for soil, six for climate and one for vegetation) with the assumption of a homogeneous soil-atmosphere column using Richards equation. Further, the behaviour of soil moisture in the upper soil zone (1 m deep or root zone) is expressed in terms of the following three independent soil parameters: effective porosity n, pore disconnectedness index c' and saturated hydraulic conductivity at soil surface K_S while storm and inter storm net soil moisture flux are coupled to storm and inter storm Probability Density Functions. The average annual evapotranspiration E_m is finally expressed as :

$$E_m = J(E_e, M_v, k_v) (E_{pa} - E_{ra}) \qquad (24)$$

$J(.)$ evapotranspiration function; E_{pa} average annual potential evapotranspiration; E_{ra} average annual surface retention; E_e exfiltration parameter as function of initial degree of saturation s_0; k_v plant coefficient. It is approximately equal to effective transpiring leaf surface per unit of vegetated land surface; M_v vegetation fraction of surface.

Further, Milly (1993) developed similar probabilistic approach for soil water storage dynamics based on Manabe model (Manabe, 1969). A key assumption is that the soil is of high infiltration capacity. The model adopts the so-called water holding capacity W_0, which is a storage capacity parameter allowing the definition of the state "reservoir is full". For well developed vegetation, W_0 is interpreted as the difference between the volumetric moisture contents θ_f of the soil at field capacity and the wilting point θ_w ($W_0 = \theta_f - \theta_w$). Furthermore, Milly (1994) adopted seasonally Poisson and exponential Probability Density Functions, together with seasonality of evapotranspiration forcing. To take into account horizontal large length scales, the spatial variability of water holding capacity W_0 was introduced, adopting a Gamma Probability Density Function with mean W_{m0}. In total, the model involved only seven parameters: a dryness index EDI = P / ETP, the mean holding capacity of soil W_{m0} and a shape parameter of the Gamma distribution,, mean storm arrival rate, and one measure of seasonality for respectively annual precipitation, potential evapotranspiration and storm arrival rate. Performing a comparison with observed annual runoff in US, it was found that the geographical distribution of calculated runoff shares at least qualitatively the large scale features of observed maps. In effect, 88% of the variance of grid runoff and 85% of the variance of grid evapotranspiration is reproduced by this model. However, it is outlined that the model presents failures within areas with elevation. Average annual precipitation and runoff over 73 large basins worldwide were also studied by (Milly and Dunne, 2002). Using precipitation and net radiation as independent variables, they compared observed mean runoff amounts to those computed by Turc-Pike and Budyko models. In northern Europe, they found a tendency for underestimation of observed evapotranspiration.

3.3 Empirical model for estimating regional evapotranspiration

Combining the water balance to the radiative balance at monthly scale, Budyko proposed an asymptotic solution in which R_n stands for average annual net radiation (which is the net energy exchange with the atmosphere equal to net radiation – sensible heat flux – latent heat flux), P average annual precipitation, E_m average (long term) annual evapotranspiration, ϕ a function expressed in Eq. (26).

$$E_m / P = \phi (R_n/P) \qquad (25)$$

$$\phi (x) = [x (tanh(x^{-1})) (1 - \cosh(x) + \sinh(x))]^{1/2} \qquad (26)$$

Where tanh(.) stands for hyperbolic tangent, cosh(.) hyperbolic cosines, sinh(.) hyperbolic sinus

According to Shiklomavov (1989) and Budyko (1974), Ol'dekop was the first to propose in 1911 an empirical formulation of the relationship between climate characteristics and water balance terms (rainfall and runoff) assuming the concept of « maximum probable evaporation» E_{max} and using the ratio P / E_{max}. According to Milly (1994), works of Budyko in 1948 resulted, on the basis of dimensional analysis, to propose the ratio R_n/P as radiative index of aridity. Conversely, the function ϕ (Eq. 26) was empirical and was derived assuming that in arid climate E_m approaches P while it approaches R_n under humid climate.Budyko model was validated using 1200 watersheds world wild computing E_m as the difference between average long term annual observed rainfall and annual observed runoff. Model accuracy is reflected by the fact that the ratio E_m /P is simulated within a relative error of 10% (Budyko, 1974). However, larger discrepancy values are found for basins with important orography. Choudhury (1999) proposed to adopt Eq. (27) to derive ϕ :

$$\phi (x) = (1+x^{-\nu})^{-1/\nu} \qquad (27)$$

where ν is a parameter depending of the basin characteristics. Milly et Dunne (2002) reported that $\nu=2.1$ closely approximates Budyko model, while $\nu=2$ corresponds to Turc-Pike model. According to Choudhury (1999), the more the basin area is large, the more ν is small and smaller is E_m. $\nu=2.6$ is recommended for micro-basins while $\nu=1.8$ for large basins. According to Milly et Dunne (2002), it was found that for a large interval of watershed areas, $\nu=1.5$ to 2.6.

Another approximation of Budyko model is the Hsuen Chun (1988) model (H.C.) introducing the ratio ID_{etp} =E_0/P and an empirical parameter k'.

$$E_m=E_0 [ID_{etp}{}^{k'} / (1+ ID_{etp}{}^{k'})]^{1/k'} \qquad (28)$$

After Hsuen Chun (1988) the value $k'=2.2$ reproduces Budyko model results. According to Pinol et al. (1991), the adjusted values of k' are in the interval 1.03 <k'< 2.40. Also, they noticed that k' depends on the type of vegetation cover. After Donohue et al. (2007), Eq. (28) may be adopted for basins with area < 1000 Km² and series of at least 5 year length.

3.4 Modeling of actual evapotranspiration for long time series and large scale applications

Simple soil water balance models based on bucket schematization have been developed to fulfil the need to simulate long time series of water balance outputs allowing the calculation of actual evapotranspiration. We focus the review on the Manabe model (1969), the

Rodriguez-Iturbe et al. (1999) model and the Bottom hole bucket model of Kobayachi et al. (2001).

3.4.1 Manabe bucket model

In fact, the single layer single bucket model of Manabe (1969) takes a central place in large scale water budget modelling. It was proposed as part of the climate and ocean circulation model. This conceptual model runs at the monthly scale and adopts the field capacity S_{FC} as key parameter. Also, it assumes an effective parameter W_k representing a fraction of the field capacity ($W_k = 0.75^* S_{FC}$). Here we notice that the field capacity S_{FC} is now expressed as a water content. The climatic forcing is represented by the potential evapotranspiration E_0. Let w be the actual soil water content. The actual evapotranspiration E_a is expressed as a linear piecewise function:

For $w \geq W_k$ $E_a = E_0$

For $w < W_k$ $E_a = E_0^*(w/W_k)$

On the other hand, the surface runoff R_s component in Manabe model depends on the actual soil moisture content in comparison to the field capacity as well as on the precipitation forcing compared to the potential evapotranspiration uptake. Let Δw the change in soil water content. Thus, surface runoff is assumed as following:

For $w = S_{FC}$ and $P > E_0$; $\Delta w = 0$ and $R_s = P - E_0$

For $w < S_{FC}$; $\Delta w = P - E_a$; $R_s = 0$

Another well-known model is FAO-56 model (Allen et al. (1998)). In fact, it is based on Manabe soil water budget. However, it takes into account the water stress through an empirical coefficient K'_s. First of all, in FAO-56 model, it is important to outline that the potential evapotranspiration is replaced by a reference evapotranspiration E_r computed using Penman-Montheith model with respect to a reference grass corresponding to an albedo value equal 0.23. Then, a seasonal crop coefficient K_c is introduced. The parameter K_c depends on both the crop type and the vegetative stage. Default K_c values are reported in (Allen et al. (1998)) for various crop types. This crop coefficient corresponds to ideal soil moisture conditions related to no water stress conditions and to good biological conditions. In real conditions, K_c is corrected by a correction coefficient K'_s ($0 < K'_s < 1$) such that the product $K_c K'_s$ includes the vegetation type as well as the water stress conditions. So actual evapotranspiration is written as:

$$E_a = K_c K'_s E_r \qquad (29)$$

According to Biggs et al. (2008) mild stress conditions would correspond to K'_s of 0.8 and moderate stress conditions to K'_s of 0.6. Based on the findings that default K_c values underestimate lysimeter experiments K_c values, Biggs et al. (2008) built a non linear regression relationships between the product ($K_c K'_s$) and the ratio of seasonal precipitation to potential evapotranspiration for various crop types. To that purpose they fitted a Beta Probability Density Function to the correction factor K'_s. They adopted lysimeter observations to fit this modified FAO-56 model.. The model explained (49–90%) of the variance in actual evapotranspiration, depending on the crop type.

3.4.2 Rodriguez-Iturbe model

In Rodriguez-Iturbe et al. (1999), the point of departure is infiltration into the soil which is expressed as function of the existing soil moisture which is reported in terms of saturation (corresponding to s= w/nZ_a where Za is effective depth of soil and n soil effective porosity). Soil drainage varies according to a power law although it is approximated by two linear segments. Consequently, it is assumed that soil drainage occurs for s exceeding a threshold value s_1, going from zero for s=s_1 to K_S for saturated condition (s=1) where K_S is the saturated hydraulic conductivity of the soil. Moreover, a saturation threshold s* is assumed to reduce evapotranspiration in case of water stress. Its value depends on the type of vegetation. Thus, for s≤s*, the evapotranspiration is computed as the potential rate scaled by the ratio s/s* while the evapotranspiration is at potential value for s> s*.

$$E_a(s)=E_0\, s/s^* \quad \text{For } s≤s^* \tag{30}$$

$$E_a(s)=E_0 \quad \text{For } s>s^* \tag{31}$$

Milly (2001) model corresponds to the case s* → 0 and K_S → infinity. According to Milly (2001), the introduction of the threshold parameter s* is much recommended especially under arid conditions. In the case where no distinction is made between forested and bare soil areas, Rodriguez-Iturbe et al. (1999) pointed out that s* is considerably lower than the field capacity S_{FC} conversely to Manabe model which corresponds to s* = 0.75 S_{FC}. Laio (2006) adopted a generalized form of Rodriguez-Iturbe et al. (1999) model by accounting for the reduction of evapotranspiration in case of water stress by introducing the soil moisture at wilting point s_w. He represented s* as a soil moisture level above which plant stomata are completely opened (Eq. 32 and Eq. 33).

$$E_a(s)=E_0\, (s-s_w)/(s^*-s_w) \quad \text{For } s≤s^* \tag{32}$$

$$E_a(s)=E_0 \quad \text{For } S_{FC} >s>s^* \tag{33}$$

On the other hand, Rodriguez-Iturbe et al. (1999) model the leakage component is represented by the exponential decay Gardner model. This model was also adopted by Guswa et al. (2002). Leakage component is assumed as exponential decay function of the effective degree of soil saturation, as well as soil characteristics (saturated hydraulic conductivity, drainage curve parameter and field capacity).

3.4.3 Bottom hole bucket model

The daily bucket with bottom hole model (BBH) proposed by Kobayashi et al. (2001) is also based on Manabe model involving one layer bucket but including gravity drainage (leakage) as well as capillary rise. Kobayashi et al. (2001) outlined that the soil moisture dynamics is better simulated by BBH than by Bucket (Manabe) model. Kobayashi et al. (2007) developed a new version of BBH named BBH-B including a second soil layer in order to take into account for the variability of the soil profile when the root zone is rather deep (1 m or more).

In the following, we focus on BBH model where forcing variables are precipitation P and potential evapotranspiration E_0. The actual evapotranspiration is assumed as:

$$E_a= M'\, E_0 \quad \text{For } s≤s^*$$
$$E_a= E_0 \quad \text{For } s>s^* \tag{34}$$

Where M' is a water stress factor updated at each time step and expressed as:

$$M'=\text{Min } (1, w/(\sigma W_{max})) \qquad \text{For } s \leq s^* \qquad (35)$$

σ: parameter representing the resistance of vegetation to evapotranspiration; $W_{max}=nZ_a$ where W_{max}: total water-holding capacity (mm); Z_a: thickness of active soil layer (mm); n: effective soil porosity.

Percolation and capillary rise term Gd(t) is assumed according to exponential function.

$$Gd(t)=\exp ((w(t)-a)/b)-c \qquad (36)$$

Where a: parameter related to the field capacity (mm); b: parameter representing the decay of soil moisture (mm); c: parameter representing the daily maximal capillary rise (mm). On the other hand, daily surface runoff Rs(t) is expressed as:

$$Rs(t)=\text{Max } [P(t)-(W_{BC}-W(t))-E_a(t)-Gd(t), 0] \qquad (37)$$

Where $W_{BC}= \eta \ W_{max}$; η : parameter representing the moisture retaining capacity ($0< \eta <1$). According to Kobayachi and al. (2001) the parameter a (which corresponds here to a/Wmax) is "nearly equal to or somewhat smaller than the field capacity". After Teshima et al. (2006), parameter b is a measure of soil moisture recession that depends on hydraulic conductivity and thickness of active soil layer Z_a. In Iwanaga et al. (2005), a sensitivity analysis of BBH model applied to an irrigated area in semi-arid region suggests that error soil moisture is most sensitive to σ, η and c.

3.5 Discussion

According to the previous presentation and model comparison, bucket type models involves one parameter in Manabe model (W_k) up to six parameters in BBH (W_{max},a,b,c,σ, η). The minimum level of model complexity for bucket type models is discussed using a daily time step by Atkinson et al. (2002). These authors introduced the permanent wilting point θ_{pwp} to refine the bucket capacity S_{bc} = $(n-\theta_{pwp})Z_a$. Also, complexity is raised by the inclusion of a separation between transpiration and evaporation from bare soil. Hence a parameter which represents the fraction of basin area covered by forests is incorporated. A linear piecewise function is assumed similarly to Rodriguez-Iturbe et al. (1999) in both cases (bare soil areas and forest areas). They suppose that storage at field capacity S_{fc} is the bucket capacity S_{bc} scaled by a threshold storage parameter fc with S_{fc} = fc S_{bc} and fc $=(\theta_{fc}- \theta_{pwp})/ (n-\theta_{pwp})$ where θ_{fc} is volumetric water content corresponding to field capacity. In addition, they assume that saturation excess runoff occurs when the storage exceeds S_{bc} and that subsurface runoff occurs when the storage exceeds S_{fc} with a piecewise non linear drainage function involving two recession parameters. These parameters are further calibrated using observed discharge recession curves while the other parameters are adapted from soil properties (via field data interpretation). Under wet, energy limited catchments authors conclude that the threshold storage parameter fc has a little control on runoff. Conversely, under drier catchments they conclude that the threshold storage parameter fc controls runoff volumes. Either, Kalma & Boulet (1998) compared simulation results of the hydrological model VIC which assumes a bucket representation including spatial variability of soil parameters to the one dimensional physically based model SiSPAT. Using soil moisture profile data for calibration, they conclude that catchment scale wetness index for very dry and very wet periods are misrepresented by SiSPAT while VIC model may better capture the water flux near and by the land surface. However, they outlined that

the difficulty of physical interpretation of the bucket VIC model parameters (maximum and minimum storage capacity) constitutes a major drawbacks of the bucket approach.
Guswa et al. (2002) also compared simulations of Richards (1D) and daily bucket model for African Savanna. They outlined that the differences between models outputs are mainly in the relationship between evapotranspiration and average root zone saturation, timing and intensity of transpiration as well as uptake separation between transpiration and evaporation. Vrugt et al. (2004) as well compared the daily Bucket model to a 3-D model (MODHMS) based on Richards equation while taking into account drainage observations. They concluded that Bucket model results are similar to MODHMS results. They also noticed that physical interpretation of MODHMS parameters is difficult since they represent effective properties. Moreover it is noticed that soil control on evapotranspiration is important in dry conditions. Besides, the introduction of a threshold parameter for evapotranspiration uptake is much recommended under arid conditions. Else, according to Rodriguez-Iturbe et al. (1999) under dry conditions, the spatial variation in soil properties has very little impact on the mean soil moisture. DeMaria et al. (2007) analyzed VIC parameter identifiability using stream flows data. Classifying four basins according to their climatic conditions (driest, dry, wet, wettest) they concluded that parameter sensitivity was more strongly dictated by climatic gradients than by changes in soil properties.

4. Rainfall runoff hydrological models

Soil water balance represents a key component of the structure of many Rainfall-runoff (R-R) models. Rainfall-runoff models are primarily tools for runoff prediction for water infrastructure sizing, water management and water quality management. On the basis of rainfall and temperature information, they aim to simulate the water balance at local and regional scales often adopting daily time step. In the majority of cases, model structure is a conceptual representation of the water balance, model parameters having to be adjusted using climatic and soil information as well as hydrological data, in order to match model outputs to observed outputs (Wagener et al., 2003). R-R models have two main components: a soil moisture-accounting module (also named production function) and a routine module (also named transfer function). In the former, the soil moisture status is up-dated while in the latter the runoff hydrograph is simulated. Models differ by the sub-models which are used for each hydrological process in both modules. The way of computing infiltration, evapotranspiration and leakage is of amount importance in the moisture-accounting module which simulates the soil moisture dynamics. It is worth noting that the Rainfall-Runoff Modelling Toolkit (RRMT), developed at Imperial College offers a generic modeling covering to the user to help him (her) to implement different lumped model structures to built his (her) own model (http://www3.imperial.ac.uk/ewre/research/software/toolkit). The system architecture of RRMT is composed by the production and transfer functions modules, and either an off-line data processing module, a visual analysis module and optimization tools module for calibration purposes (Wagener et al. 2001). In this section, we focus on evapotranspiration sub-models of two well-used R-R models (section 4.1). Then, we review the main steps of the calibration process required to estimate the model parameters (section 4.2). Finally three case studies are reported (section 4.3).

4.1 Evapotranspiration sub models

Despite the focus on runoff results in R-R modeling, evapotranspiration computation is a key part of R-R models. As an example, we emphasize the evapotranspiration sub-model of

GR4 model which is a parsimonious lumped model proposed by CEMAGREF (France) and running at the daily step with four parameters. A full model description is available in (Perrin et al., 2003). At each time step, a balance of daily rainfall and daily potential evapotranspiration is performed. Consequently, a net evapotranspiration capacity E_n and a net rainfall P_n are computed. If $P_n \neq 0$, a part P_s of P_n fills up the soil reservoir (so, P_s represents infiltration). It is noticeable that this quantity P_s depends on the actual soil moisture content w according to a non linear decreasing function of the w/x_1 where x_1 is the maximum capacity of the reservoir soil (which might represent the field capacity). On the other hand, if the net evapotranspiration capacity $E_n \neq 0$, actual evapotranspiration E_s is computed as a non linear increasing function of the water content involving the ratio w/x_1. Also, this function is parameterized through the ratio E_n/x_1 which refers to the characteristics of climate-soil complex. Furthermore, a leakage component is assumed with a power law function of the reservoir water content w.

$$\text{For } P \geq E_0; \quad P_n = P - E_0 \quad \text{and} \quad E_n = 0 \tag{38}$$

$$\text{For } P < E_0; \quad P_n = 0 \quad \text{and} \quad E_n = E_0 - P \tag{39}$$

$$E_s = w \, (2-(w/x_1)) \tanh(E_n/x_1)/\{1+[(1-wx_1)\tanh(E_n/x_1)]\} \tag{40}$$

Where tanh(.) stands for hyperbolic tangent.

As second example, we underline the sub-models adopted in the *HBV* conceptual semi-distributed model proposed by the Swedish hydrological institute (Begström, 1976). The fraction ΔQ of precipitation entering the soil reservoir is assumed as power law function of the ratio (w/FC) of reservoir water content w to a parameter FC representing soil field capacity in HBV model.

$$\Delta Q = P_e[1-(w/FC)^{\beta'}] \tag{41}$$

Where β' is a calibration parameter usually estimated by fitting observed and simulated runoff data. Also, P_e is effective precipitation. In addition, the actual evapotranspiration is a piecewise linear function. The control of actual evapotranspiration rates is performed using a parameter PWP representing a threshold water content. If w< PWP, the evapotranspiration uptake is a fraction of the potential evapotranspiration E_0 otherwise it is at potential rate.

$$E_a/E_0 = w/PWP \text{ for } w<PWP;$$

$$\text{and } E_a = E_0 \text{ for } w>PWP \tag{42}$$

4.2 Model calibration issues

As runoff has been for long time the main targeted response of rainfall-runoff modeling, rainfall-runoff models were often adjusted according to runoff observations. So far, observations from other control variables such as soil moisture content (Lamb et al., 1998), water table levels (Seibert, 2000) and either low flows (Dunne, 1999) have been adopted to enhance runoff predictions. Calibration of model parameters against runoff data is often performed using criteria such as bias and Root Mean Square Error (RMSE), which helps quantifying the discrepancy between observed discharges y_0 and simulated discharges y_i over a fixed time period with N observations.

$$RMSE = \left(\frac{1}{N} \sum_{i=1}^{i=N} (y_{si} - y_{oi})^2 \right)^{1/2} \tag{43}$$

The difficulty in the calibration process is that various parameter sets and even model structures might result in similarly good levels of performance, which constitutes a source of ambiguity as out pointed by Wagener et al. (2003) and many other authors before them (see the literature review of Wagener et al. (2003)). Also, it is noticeable that this problem of ability of various model structures and model parameters to perform equal quality with respect to matching observations is not dependent of the calibration process itself. In other words, the use of a performing optimisation tool does not prevent the problem. Another question is related to the single versus multi objective optimization. Wagener et al. (2003) reported that "single objective function is sufficient to identify only between three and five parameters" while lumped R-R models usually adopt far superior number of parameters. Multi-objective approach of calibration using additional output variables such as water table levels or soil moisture observations has been introduced to deal with the problem. Yet, inadequate model structure may be responsible of mismatching between observed and simulated outputs, as related by Boyle et al. (2000).

4.3 Case studies
Three case studies are presented in this section. In the first case, we propose a method for calibrating the empirical parameter k' of Hsuen Chun (1988) (Eq. 28). In the second case, we propose as example of calibrating HBV model using both runoff data and regional evapotranspiration information. In the third case, calibration of BBH model is performed using both runoff data and regional evapotranspiration information.

4.3.1 Fitting empirical models of regional evapotranspiration
This case study is presented in Bargaoui et al. (2008) and Bargaoui & Houcine (2010). It is aimed to calibrate the H.C. model using climatic, rainfall and runoff data from gauged watersheds. Monthly temperature and solar radiation data as well as annual rainfall and runoff data from various locations in Tunisia listed in Table 1 are adopted to calibrate the parameter k' of the empirical Hsuen Chen model (Eq. 28). To this end, 18 rainfall stations and 20 river discharge stations are considered, as well as 8 meteorological stations (Table 1). On the other hand, the potential evapotranspiration E_0 is computed at monthly scale using Turc formula.

$$E_0 = 0.4\, T_m\, [(R_g/N_j)+50]\, /\, [Rg+15] \tag{44}$$

T_m : monthly average temperature in (°C); R_g : global solar radiation (cal.cm^{-2} month^{-1}); N_j : number of days by month

For each river basin, simulated average (long term) annual evapotranspiration is computed using Eq. (28). Then, simulated mean annual runoff is computed as the difference between observed mean annual precipitation and simulated average annual evapotranspiration. The fitting of annual simulated runoff to annual observed runoff using the 20 river discharge stations results in k'= 1.5. The good adequacy of the model is well reflected in the plot of average simulated versus average observed annual runoff (Fig. 1).

River discharge stations			Rainfall stations			Meteorological stations		
Stations	Latitude	Longitude	Stations	Latitude	Longitude	Stations	Latitude	Longitude
Jebel Antra	36°57′18″	9°27′45″	Ouchtata	36°57′53″	8°60′1″	Sfax	34°43′0″	10°41′0″
Joumine Mateur	37°2′19″	9°40′56″	Cherfech	36°57′0″	10°3′13″	Tunis	36°51′0″	10°20′0″
Zouara	36°54′15″	9°7′1″	Tabarka	36°56′59″	8°44′50″	Tabarka	36°57′0″	8°45′0″
Barbara	36°40′32″	8°32′56″	El Kef	36°10′53″	8°42′57″	Bizerte	37°14′0″	9°52′0″
Rarai sup.	36°27′36″	8°21′20″	Mellègue	36°7′16″	8°30′2″	Jendouba	36°29′0″	8°48′0″
Mellegue K13	36°7′1″	8°29′52″	Tajerouine	36°27′32″	9°14′57″	El Kef	36°8′0″	8°42′0″
Mellegue Rmel	36°1′1″	8°37′14″	Mejez El Bab	36°39′3″	9°36′17″	Kairouan	35°4′0″	10°4′0″
Haffouz	35°37′58″	9°39′33″	Tunis	36°47′23″	10°10′23″	Siliana	36°4′0″	9°22′0″
Merguellil Skhira	35°44′24″	9°23′3″	Feriana	34°56′49″	8°34′29″			
Chaffar	34°33′49″	10°29′14″	Jendouba	36°30′14″	8°46′52″			
Joumine Tine	36°58′3″	9°43′2″	Sejnane BV	37°3′35″	9°14′46″			
Miliane, Tuburbo Majus	36°23′39″	9°54′43″	Ksour	36°45′22″	9°28′27″			
M′khachbia aval	36°43′22″	9°24′24″	Sers	36°4′19″	9°1′25″			
Haidra Sidi Abdelhak	35°56′59″	8°16′22″	Ghardimaou	36°27′2″	8°25′58″			
Medjerda Jendouba	36°30′40″	8°46′7″	Bou Salem	36°36′30″	8°57′57″			
Sejnane	37°11′37″	9°30′16″	Merguellil H.	35°38′8″	9°40′36″			
Tessa Sidi Medien	36°16′44″	8°57′14″	Merguellil Skhira	35°44′24″	9°23′3″			
Rarai plaine	36°29′16″	8°32′18″	Chaffar PVF	34°40′0″	10°5′0″			
Ghezala-Ichkeul	37°4′35″	9°32′12″						
Douimis	37°12′50″	9°37′38″						

Table 1. Location of stations to calibrate H.C. model (after Bargaoui &Houcine, 2010)

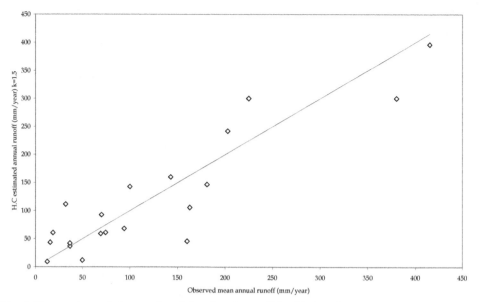

Fig. 1. Comparison of observed and simulated runoff for 20 river basins

4.3.2 Multicriteria calibration of HBV model using regional evapotranspiration information

This application is presented in Bargaoui et al. (2008). The idea is to use the information about the climatic regime as a driver for runoff prediction. Effectively, for a large number of basins with areas in the interval 50 à 1000 km², Wagener et al., (2007) suggested that there is a significant correlation between annual runoff and the ratio of forcing variables P/E_0. In the same way, we seek to use information about average (regional) actual evapotranspiration which is a bio-climatic indicator as means to improve accuracy of runoff predictions. To develop these ideas, the HBV rainfall-runoff model was adopted, coupled to a SCE-UA optimization tool. The calibration method adopts an objective function combining three criteria: minimisation of runoff root mean square error, minimisation of water budget simulation error, minimisation of the difference between mean annual simulated evapotranspiration E_a and regional E_m. The case study is a mountainous watershed of Wadi Sejnane (Tunisia). Mean daily runoff observations from September 1964 to August 1969 are available for a hydrometric station controlling an area of 378 km². Average basin annual rainfall is 931 mm/year. Over 8 years of rainfall observations, the minimum value of the series is 628 mm/year while the maximum value is 1141 mm/year denoting an important rainfall inter annual variability. Mean annual discharge is 2.43 m³/s. Average evapotranspiration computed using HC model (Eq. 28) with k'=1.5 results in E_m=643 mm/year. To calibrate the HBV model parameters, we adopt the period 1964/1967 for calibration and the period 1967/1969 for validation. The minimization of the objective function is performed using SCE-UA algorithm (Duan et al., 1994) in order to adjust 10 parameters (while 7 other HBV parameters have been set constant because they were found insensitive). First, the Nash coefficient of mean daily discharges is chosen as objective function F_0=Nash$_R$. The resulting value F_0=0.81 is quite good. However, for the validation period the ensuing optimal parameter set results in very poor fitting with a negative value of the Nash coefficient ($Nash_R$ = -0.084). Consequently, the objective function was modified to F_1 integrating the average model error (bias) of runoff output. Hence,

$$F_1 = Nash_R - w' ER_{RA} \tag{45}$$

Where ER_{RA} is the absolute relative error with respect to annual discharge. The weight coefficient w' = 0.1 is adopted according to Lindström and al. (1997) and helps aggregate the two criteria $Nash_R$ and ER_{RA}. In fact, the adoption of ER_{RA} aims to consider climatic zonality during the calibration process. Resulting optimal solution corresponds to $Nash_R$ =0.81 and ER_{RA} = 5%, which is believed good performance. It is worth noting that this modification of the objective function greatly improved $Nash_R$ also for the validation period ($Nash_R$ =0.55). The mean annual simulated evapotranspiration using HBV model is equal to 728 mm/ year while the H.C. model with k'=1.5 results in 643mm/year. To try to overcome such overestimation, it was proposed to directly include the information about evapotranspiration by adopting a new objective function F_2.

$$F_2 = Nash_R - 0,1 ER_{RA} - 0,1 ER_{ETRG} \tag{46}$$

Where ER_{ETRG} is the absolute relative error with respect to mean annual evapotranspiration (simulated by HBV versus estimated by H.C with k'=1.5). The resulting runoff Nash is a little smaller ($Nash_R$ =0.79) than for F_1, but a real improvement is obtained during the validation period ($Nash_R$ = 0.68). Fig. 2 reports HBV estimated annual evapotranspiration obtained with the optimal HBV solution (squares) versus annual rainfall. Comparatively, we also report annual evapotranspiration as evaluated using H.C model with k'=1.5 (interrupted line). Effect of year to year rainfall fluctuation on HBV estimations is well seen in the graph.

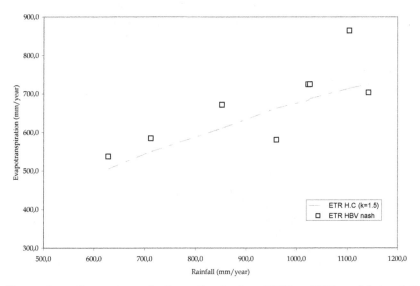

Fig. 2. Comparison of evapotranspiration estimates from HBV and HC models in relation with rainfall

4.3.3 Multicriteria calibration of BBH model using regional evapotranspiration information

In the third application it is aimed to compare BBH model results using the decadal time step. A part of this case study is presented in Bargaoui & Houcine (2011) using monthly data for model evaluation. Here will report results of decadal evaluations. Data are from the Wadi Chaffar watershed (250 km²) situated under arid climate, South Tunisia. Vegetation cover comprises mainly olives. Meteorological data (solar radiation, air temperature and humidity, sky cloudiness, wind speed and Piche evaporation) are available from September 1989 to August 1999 for computing the daily reference evapotranspiration E_0 according to Allen et al. (1998). E_0 is multiplied by the crop coefficient K_c of olives trees to obtain daily potential evapotranspiration (Allen et al., 1998). Daily average basin rainfalls are available from September 1985 to August 1999. Stream discharge data are available for the basin outlet at the daily time step from September 1985 to August 1999. In the period September 1985 to August 1989, meteorological data are missing and the used E_0 values are the daily long term average computed for September 1989- August 1999. The H.C. model results in an average annual evapotranspiration E_m = 213 mm/year (Bargaoui & Houcine, 2010). BBH model inputs are precipitation and potential evapotranspiration and seven parameters are to be calibrated. To reduce the number of calibrated parameters, we first fix the thickness of active soil layer Z_a (in mm) and the effective soil porosity n (unit less). Also, we undertake a reformulation of leakage component L(s) by using the model of Guswa et al. (2002) where

$$L(s) = K_S \frac{e^{B(s-S_{FC})} - 1}{e^{B(1-S_{FC})} - 1} \tag{47}$$

where s is the degree of saturation (unit less); K_S saturated hydraulic conductivity at soil surface (mm/day); B is the soil water retention curve shape parameter; S_{FC} (unit less) is the field capacity; $W_{max} = nZ_a$ (W_{max} is the total water-holding capacity in mm).

Coupling this expression with pedo-transfer functions it makes it possible after Bargaoui & Houcine (2010), to derive the parameters (a, b, c) as following using pedo-transfer parameters K_S , B and S_{FC}:

$$a = W_{max} \left[S_{FC} - \frac{1}{B} Ln \left(K_S \frac{1}{e^{B(1-S_{FC})} - 1} \right) \right] \tag{48}$$

$$b = W_{max} \frac{1}{B} \tag{49}$$

$$c = \left(\frac{1}{e^{B(1-S_{FC})} - 1} \right) K_S \tag{50}$$

In this case, the model by Rawls et al. (1982) is adopted for K_S estimation while S_{FC} is derived according to the Cosby (1984) and Saxton et al. (1986) models recently adopted by Zhan et al., (2008). Finally $B = 9$ is assumed in agreement with Rodriguez-Iturbe et al. (1999). The dominant soil type is considered to represent the soil characteristics. So, the value n=0.34 corresponding to a sandy soil was adopted; these assumptions result in K_S = 3634 mm/d and S_{FC}= 0.166. Also, after many trials the value Z_a= 0.5 m was adopted. The two remaining parameters σ and η (0< σ <1; 0< η <1) represent respectively the resistance of vegetation to evapotranspiration and the moisture retaining capacity. The problem is now to fit the parameters σ and η. They are adjusted using two different methods: i.e. using only observed runoff (method 1) and using both observed runoff and regional evapotranspiration information (method 2). Also BBH model has been completed adopting a , contributing area sub-model (Betson, 1964); Dunne et Black (1970). According to this assumption, runoff originates from a part of the watershed (contributing area) contrarily to the assumption of runoff occurring from the entire watershed. For a fixed day j, the contributing area CA_j is herein assumed linked to the soil moisture content according to Dickinson & Whiteley (1969). Additionally, a logistic Probability Density Function as a function of humidity index IH_j is adopted with parameters a_c and b_c (Eq. 51). It means that the mean contributing area is a_c and that the variance of the contributing area is $(b_c\pi)^2/3$. The humidity index takes account for the rainfall accumulated during the actual day and the IX previous days (Eq. 52).

$$CA_j = \frac{e^{((IH_j - a_c)/b_c)}}{(1 + e^{((IH_j - a_c)/b_c)})} \tag{51}$$

$$IH_j = W_{j-1} + \omega'' \sum_{l=0}^{IX} P_{j-l} \tag{52}$$

where ω'' is a fixed weight ($\omega'' = 0.1$). Then, two cases are considered: case (a) when the total basin area contributes to runoff at the basin outlet; case (b) when only a contributive area gives rise to runoff at the outlet.

After many trials and errors we assumed IX= 90 days, $a_c = 20$ and $b_c = 10$ in case (b). The model was calibrated for σ and η using daily hydro meteorological data (solar radiation, air temperature, air humidity, mean areal rainfall) as well as daily runoff records and also average annual evapotranspiration. The decadal, monthly and annual totals are adopted to evaluate model performance.

In each case (a) and (b), a first criterion based on the matching of decadal runoff (Eq. 53) is adopted to delineate adequate solutions for σ and η ($0 < \sigma < 1$; $0 < \eta < 1$). A supplementary criterion is based on the matching of long term annual evapotranspiration (Eq. 54).

$$C_y(\sigma, \eta) = \frac{1}{N} \sum_{i=1}^{N} |(y_{si} - y_{oi})/y_{oi}| \tag{53}$$

$$C_E(\sigma, \eta) = \frac{1}{N'} \sum_{i=1}^{N'} |(E_{si} - E_m)/E_m| \tag{54}$$

In Eq. (53), y_{oi} and y_{si} are respectively decadal observed and simulated volume runoff and N is the number of simulated decades. In Eq. (54), E_{si} is simulated annual evapotranspiration and N' is the number of simulated years.

For each pair of simulated (σ, η) ($0 < \sigma < 1$; $0 < \eta < 1$), candidate solutions verifying the criterion $C_y(\sigma, \eta) < \alpha$ (Eq. 53) with $\alpha = 20\%$ the Nash coefficient R_N is then evaluated. Pairs for which it is found that $R_N > 0.5$, are thus selected. Also, introducing E_m for calibration method 2, the absolute value $C_E(\sigma, \eta)$ of the relative error between mean annual simulated evapotranspiration and E_m, is used through the additional selection criterion of (Eq. 54).

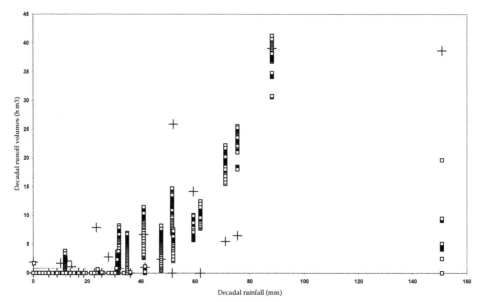

Fig. 3. Estimated decadal runoff versus decadal precipitation with the assumption of total watershed contributing to runoff (+ represent observed volumes and squares represent simulated volume for the selected pairs of (σ, η))

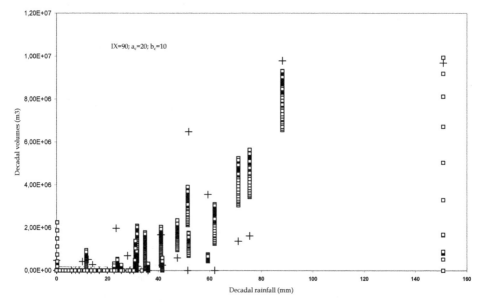

Fig. 4. Estimated decadal runoff versus decadal precipitation with the assumption of contributive area (+ represent observed volumes and squares represent simulated volume for the selected pairs of (σ, η)).

Pairs of simulated (σ, η) $(0 < \sigma < 1; 0 < \eta < 1)$ which satisfy both $C_y(\sigma, \eta) < 20\%$; $R_N > 0.5$ and $C_E(\sigma, \eta) < \alpha'$ with $\alpha' = 30\%$ are finally selected as adequate solutions.

Fig. 3 and 4 report model outputs for sets of (σ, η) fulfilling the above conditions under the assumptions of cases (a) and (b) in case where E_m information is included. Estimated decadal volumes (squares) for the selected pairs of (σ, η) are compared to observed decadal volumes (+) and are reported versus precipitation data. Fig. 3 is related to case (a) corresponding to the assumption of total watershed contributing to runoff. Fig. 4 is related to case (b) assuming a contributing area. The results suggest that the introduction of contributing area outcomes produce outputs which result in a better fitting of the rainfall-runoff evolution. In effect, in the case of total area contributing no solution is found able to simulate the most rainy decade,(squares are far from the symbol + for the Rainiest decade). Conversely, some solutions are found able to reproduce the most rainy decade if we consider contributing area scheme (some squares are located near the +). Also, evapotranspiration information has greatly reduced the interval of acceptable solutions. Effectively, selected solutions are such that $0.15 < \sigma < 0.35$ and $0.15 < \eta < 0.25$.

5. Conclusions

The simulation of evapotranspiration using the water balance equation is part of hydrological modelling (rainfall-runoff models) and is also important in the framework of global circulation models (Land surface models). A lot of models are now functioning and their formulation is based on different assumptions on soil characteristics in relation with soil moisture, transpiration schemes, as well as infiltration and runoff schemes.

Empirical models for estimating regional evapotranspiration are worth noting for estimating average long term evapotranspiration. They are generally based on climatic information (rainfall and potential evapotranspiration). They often require the adjustment of a single empirical parameter. Under particular climate and soil vegetation, evapotranspiration is controlled by soil moisture dynamics. Thus, Bucket type soil water budget models are worth noting for estimating time series of actual evapotranspiration at smaller time scales (daily to monthly). They involve from one parameter such as in the Manabe model (with parameter W_k) up to six parameters such as in BBH model (with parameters $W_{max}, a, b, c, \sigma, \eta$). Parameters are linked to soil, climatic and vegetation characteristics. However, it is generally believed that the temporal variability of soil moisture series is mostly dependent on the rainfall variability especially under conditions of low precipitations. On the other hand, soil parameters such as field capacity, hydraulic conductivity at saturation and wilting point potential are key parameters controlling the evapotranspiration model outputs. One way to derive soil parameters is to adopt pedo transfer functions. Transpiration which corresponds to vegetation uptake is regulated by stomata and driven by atmospheric demand. It is widely represented by a linear piecewise function with parameters depending on vegetation characteristics. Thus, in computing evapotranspiration, a main assumption is the linear piecewise function of evapotranspiration in relation with potential evapotranspiration for taking account for soil water stress. Such an assumption is underlined in several rainfall runoff models (for example the two models GR4 and HBV studied here adopt such analytical form). Model adequacies introduce the question of the choice of the objective function as well as the output variables adopted for model evaluation. In the case studies presented here, results suggest that the introduction of the information about average (long term) annual evapotranspiration may help improving the accuracy of the water balance simulation results. In effect the runoff Nash coefficient is found to be improved during the validation period in the case where long term evapotranspiration is taking account during the calibration process.

6. Annexe

6.1 Glossary
a: parameter related to the field capacity (mm)
a' : pore size distribution parameter
a_c: Logistic density distribution parameter
$\alpha(\theta)$: the root efficiency function.
$A(t)$: the fraction of area for which the infiltration capacity is less than $i(t)$
B : the soil water retention curve shape parameter;
b: parameter representing the decay of soil moisture (mm);
b'' is the infiltration shape parameter.
b_c: Logistic density distribution parameter
β: surface slope angle
β' : a calibration parameter in HBV model
c' : pore disconnectedness index
c: parameter representing the daily maximal capillary rise (mm)
$C_\%$: soil percent of clay
CA_j: : the contributing area
cos: the cosinus function
d_1 : depth of near surface soil layer
d_2 : depth of root zone soil layer;
D:soil-water diffusivity parameter
$\Delta\psi$: difference in average matrix potential before and after wetting
$\Delta\theta$: difference in average soil water content before and after wetting
Δw : the change in soil water content
Δz : soil depth.
C_1: parameter,
C_2: parameter,
$E(s,t)$: : evapotranspiration
E_a : actual evapotranspiration
EDI : dryness index
E_e exfiltration parameter as function of initial degree of saturation s_0
E_g : bare soil evaporation rate at the surface
E_m average annual evapotranspiration
E_n: net evapotranspiration capacity
E_r: reference evapotranspiration according to FAO model
E_{tr} : transpiration rate from the root zone of depth d_2
ER_{ETRG} : the absolute relative error with respect to mean annual evapotranspiration
E_{pa} average annual potential evapotranspiration

ER_{RA} : the absolute relative error with respect to annual discharge

E_{Turc} : monthly potential evapotranspiration (mm);

E_{ra} average annual surface retention

f : infiltration capacity

F cumulative infiltration for a rainfall event

fc: threshold storage parameter

FC: representing soil field capacity in HBV model

Gd(t): Daily percolation and capillary rise term

$g_r(\theta,z)$: vegetation uptake of soil moisture

I(s,t): infiltration into the soil

$I_{nf}(\theta,z_0)$: precipitation infiltrating into the soil

i(t): infiltration capacity at time t.

i_{max} : maximum value of infiltration capacity

Ω_f (t) : the cumulative infiltration

J(.): evapotranspiration function

k : intrinsic permeability

k(1): intrinsic permeability at saturation

K: hydraulic conductivity

K (1) hydraulic conductivity at saturation

k_v : plant coefficient

K_{av} : average saturated hydraulic conductivity

k' : parameter of HC model

K_c : crop coefficient

K_S : the saturated hydraulic conductivity;

K'_s : correction coefficient of the crop coefficient

κ: shape parameter of the Gamma distribution

l: factor linked to soil matrix tortuosity

L(s,t) :leakage

LAI : Leaf area index

λ: mean storm arrival rate

M_v : vegetation fraction of surface.

μ : dynamic viscosity of water;

n: soil effective porosity

ν : parameter

θ: volumetric water content

θ_f :the volumetric moisture contents of the soil at field capacity

θ_w: the volumetric moisture contents at wilting point

θ_{pwp}: permanent wilting point

θ_g : volumetric water contents of near surface soil layer;

θ_s : saturated soil moisture content

θ_2 : volumetric water contents of root zone soil layer;

θ_{geq} : equilibrium surface volumetric soil moisture content

θ_1 : specific value of soil moisture content

θ_0 : specific value of soil moiqture content

N: number of observations

N_j: number of days by month

$Nash_R$: Nash coefficient of mean daily discharges

P : average annual precipitation

P_e : effective precipitation

PWP : parameter representing a threshold water content in HBV model.

P_g : precipitation infiltrating into the soil;

P_n: net rainfall

q_2 : rate of drainage out of the bottom of the root zone;

R_1 : (s cm^{-1}) a resistance to moisture flow in soil

R_2 : (s cm^{-1}) is vegetation resistance to moisture flow;

R_n : average annual net radiation

R_s : surface runoff

R_g : global solar radiation (cal.cm^{-2} month^{-1})

$r(z)$: a root density function (cm^{-1})

ρ_w : density of the water;

s: relative soil moisture content or degree of saturation

s*: saturation threshold

s_1 :threshold value of soil saturation

s_w: soil moisture at wilting point.

s_0: initial degree of saturation

S : sorptivity

$S_\%$: soil percent of sand

S_{bc}: bucket capacity

S_{FC} : soil field capacity

S_{fc}: storage at field capacity

σ : parameter representing the resistance of vegetation to evapotranspiration;

t: time

T_m : monthly average temperature in (°C);

$u(z,t)$: local transpiration uptake

w : the actual soil water storage

w_0: critical soil water storage in Budyko model

W_0: water holding capacity

W_k : a fraction of the soil field capacity

W_{max} : total water-holding capacity (mm);

W_{m0}: mean water holding capacity

ω : a fixed weight

ψ : soil moisture potential (bars)

ψ_p : leaf moisture potential (bars)

$\psi*$: the wilting point potential

$\psi(1)$: the bubbling pressure head which represents matrix potential at saturation.

x_1 : maximum capacity of the reservoir soil

y_i: simulated discharges

y_0: observed discharges

z: the vertical coordinate (z>0 downward from surface)

Z_a: thickness of active soil layer (mm);

z_0 : the vertical coordinate at the surface

7. References

Allen RG, Pereira LS, Raes D, Smith M. (1998). Crop evapotranspiration. Guidelines for computing crop water requirements. *FAO Irrigation and drainage*. Paper No. 56. FAO, Rome.

Atkinson, S. E., R. A. Woods, and M. Sivapalan (2002), Climate and landscape controls on water balance model complexity over changing timescales, *Water Resour. Res.*, 38(12), 1314, doi:10.1029/2002WR001487.

Bargaoui Z., Dakhlaoui H, Houcine A. (2008). Modélisation pluie-débit et classification hydroclimatique. *Revue des sciences de l'eau*. 21 (2) : 233-245.

Bargaoui Z., Houcine A. (2010) Sensitivity to calibration data of simulated soil moisture related drought indices, *Revue Sécheresse*, 21(4), 1-7.

Bargaoui Z., Houcine A. (2011) Calibration of an evapotranspiration model using runoff records and regional evapotranspiration, *in Hydro-climatology: Variability and change IAHS Pub. 344*, 21-26.

Begström S. (1976). Development and application of a conceptual rainfall-runoff model for the scandanivain catchment , SMHI RH07, Norrköping.

Betson R. P. (1964). What is watershed runoff? *Journal of geophysical research*. 69(8). 1541-1552. In: Streamflow generation processes. Selection, Introduction and commentary by K. J. Beven. Benchmark Papers in Hydrology. IAHS Press 2006.

Biggs, T.W. , P. K. Mishra, H. Turral. (2008). Evapotranspiration and regional probabilities of soil moisture stress in rainfed crops, southern India, Agric. *Forest Meteorology*, doi:10.1016/j.agrformet.2008.05.012

Boulet G., Chehbouni A., Braud I., Vauclin M., Haverkamp R., Zammit C. (2000). A simple water and energy balance model designed for regionalization and remote sensing data utilization. *Agricultural and Forest Meteorology* 105 (2000) 117–132

Boyle, D.P., Gupta, H.V., Sorooshian, S., (2000). Toward improved calibration of hydrologic models: Combining the strengths of manual and automatic methods. Water R. Res. 36 (12), 3663–3674.

Braud I., Dantas Antonino A.C., Vauclin M., Thony J.-L., Ruelle P., A (1995). Simple Soil Plant Atmosphere Transfer model (SiSPAT) development and field verification, *J. Hydrol.* 166 (1995) 213–250.

Brooks, R.H., Corey, A.T., (1964). Hydraulic properties of porous media, *Hydrology paper, 3*. Colorado State University, Fort Collins.

Budyko, M. I. (1974) Climate and life. Academic Press. 508p.

Ceballos A., Martınez-Fernandez J.,, Santos F.& Alonso P. (2002). Soil-water behaviour of sandy soils under semi-arid conditions in the Duero Basin (Spain). *Journal of Arid Environments* (2002) 51: 501–519. doi:10.1006/jare.2002.0973, available online at http://www.idealibrary.com on

Childs, E.C. and Collis-George. N., (1950). The permeability of porous materials. *Proc. R. Soc.,London*, 201A: 392 405.

Choudhury B.J. (1999). Evaluation of an empirical equation for annual evaporation using field observations and results from a biophysical model. *J. Hydrol.* 216. 99-110.

Clapp, R. B., and G. M. Hornberger (1978). Empirical equations for some soil hydraulic properties, *Water Resour. Res.*, 14, 601–604.

Cosby (1984) A statistical exploration of the relationship of soil moisture characteristics to the physical properties of soils. *Water Resour. Res* 20(6). 682-690.

Cosby BJ, Hornberger GM, Clapp RB, Ginn TR. (1984). A Statistical Exploration of the Relationships of Soil Moisture Characteristics to the Physical Properties of Soils. *Water Resour. Res.* 20 (6) : 682-690.

DeMaria, E. M., Nijssen, B., Wagener, T. (2007) Monte Carlo sensitivity analysis of land surface parameters using the Variable Infiltration Capacity model, *Journal of geophysical research*, 112, D11113, doi:10.1029/2006JD007534, 2007

Dickinson., W.T., Whiteley, H. (1969) Watershed areas contributing to runoff, *Hydrological sciences Journal*. 12-26. Special issue: rainfall runoff modelling.

Dickinson R.E., Henderson-Sellers A., Kennedy P.J., Wilson M.F. (1986). Biosphere-Atmosphere Transfer Scheme (BATS) for NCAR Community Climate Model . In: Evaporation. Selection, Introduction and commentary by J. H. C. Gash and W.J. Shuttleworth. Benchmark Papers in Hydrology. ISSN 1993-4572. IAHS Press 2007.

Donohue R.J., Roderick M. L. and T.R. McVicar (2007). On the importance of including vegetation dynamics in Budyko hydrological model. *Hydrol. Earth Syst. Sci.*, 11, 983-995.

Duan, Q., Sorooshian, S. and Gupta, V. ,(1994). Optimal Use of the SCE-UA Global Optimisation Method for Calibrating Watershed Models. *Journal of Hydrology*, No. 158, 265-284.

Dunne, T., Black, R.D., (1970). Partial area contributions to storm runoff in a small New England watershed. *Water Resour. Res* 6, 1296-1311. In: Streamflow generation processes. Selection, Introduction and commentary by K. J. Beven. Benchmark Papers in Hydrology. IAHS Press 2006.

Dunne S. M. (1999). Imposing constraints on parameter values of a conceptual hydrological model using base flow response. *Hydrology and earth system sciences* 3: 271-284.

Eagleson PS. Climate, Soil, and Vegetation. (1978 a). Dynamics of the Annual Water Balance. *Water Resour. Res* 14 (5) : 749-764

Eagleson P. (1978 b). Climate, Soil, and Vegetation. The Expected Value of Annual Evapotranspiration. *Water Resour. Res* . 14, 5. 731- 739.

Entekhabi, D., Eagleson, P.S., (1989). Land surface fluxes hydrology parameterization for atmospheric general circulation models including subgrid variability. *J. Climate* 2 (8), 816-831.

FAO (1980). Drainage design factors. *Irrigation and Drainage* Paper No. 38. Rome. 52 pp

Feddes RA, Hoff H, Bruen M, Dawson T, de Rosnay P, Dirmeyer P,. (2001). Modeling root water uptake in hydrological and climate models. *Bull Amer Meteorol Soc.* ;82(12):2797-809.

Franks, S.W., Beven, K.J., Quinn, P.F., Wright, I.R., (1997). On the sensitivity of the soil-vegetation-atmosphere transfer (SVAT) schemes: equifinality and the problem of robust calibration. *Agricultural and Forest Meteorology* 86, 63-75.

Gardner, W . R., (1958). Somes teady-state solutions of the unsaturated moisture flow equation with application to evaporation from a water table, *Soil Sci.*, 85(4), 228-232, 1958.

Gardner , C. M. K., Robinson D. A., Blyth K. , Cooper J.D. (2001). Soil water content. In: Soil and environmental analysis: physical methods (Second Edn) (Ed. By K. A. Smith and C. E. Mullins), 1-64. Marcel Dekker Inc., New York. USA.

Green, W.H., Ampt, G.A., (1911). Studies on soil physics. I. Flow of air and water through soils. *J. Agric. Sci.* 4, 1–24.

Guswa A.J. (2005). Soil-moisture limits on plant uptake: An upscaled relationship for water-limited ecosystems. *Advances in Water Resources* 28. 543–552

Guswa, A.J., Celia, M, Rodriguez-Iturbe, I. (2002) Models of soil moisture dynamics in ecohydrology: A comparative study. *Water Resour. Res* ., 38 (9), 1166-1180.

Hsuen-Chun Y. (1988). A composite method for estimating annual actual evapotranspiration. *Hydrol. Sci. J.* ; 33 (4, 8): 345 - 356.

Hofius K. (2008). Evolving role of WMO in hydrology and water resources management. *WMO Bulletin* 57(3).147-151.

Iwanaga R, Kobayashi T, Wang W, He W, Teshima J, Cho H. (2005). Evaluating the Irrigation requirement at a Cornfield in the Yellow River Basin Based on the "Dynamic Field Capacity " *J. Japan Soc. Hydrol. & Water resour.*; 18: 664-674.

Jackson R. B. , J. S. Sperry, and T. E. Dawson, (2000). Root water uptake and transport: Using physiological processes in global predictions. *Trends Plant Sci.*, 5, 482–488.

Jeffrey P. Walker*, Garry R. Willgoose1, Jetse D. Kalma (2004) . In situ measurement of soil moisture: a comparison of techniques *Journal of Hydrology* 293. 85–99

Johnson K. D., Entekhabi D., Eagleson P.S. (1993). The implementation and validation of improved land surface hydrology in an atmospheric general circulation model. *Journal of Climate* 6. 1009-1026.

Kalma, J.D. , Boulet, G. (1998) Measurement and prediction of soil moisture in a medium-sized catchment. *Hydrological Sciences Journal*, 43(4), Special issue: onitoring and Modelling of Soil Moisture: Integration over Time and Space, 597-610.

Kobayashi T, Matsuda S, Nagai H, Teshima J. (2001). A bucket with a bottom hole (BBH) model of soil hydrology. In: Soil–Vegetation–Atmosphere Transfer Schemes and Large–Scale Hydrological Models (ed. Dolman, H. et al.), IAHS Publication. 270. 41-45.

Kobayashi T., Teshima J., Iwanaga R., Ikegami D., Yasutake D. , He W., Cho H.. (2007) An improvement in the BBH model for estimating evapotranspiration from cornfields in the upper Yellow river. *J. Agric. Meteorol.* 63(1). 1-10.

Lai C-T, Katul G. (2000).The dynamic role of root-water uptake in coupling potential to actual Transpiration. *Advances in Water Resources* 23 . 427±439

Laio F. (2006). A vertically extended stochastic model of soil moisture in the root zone. *Water Resour. Res* ., 42, W02406, doi:10.1029/2005WR004502,

Lamb R., K. Beven et S. Myrabo (1998) : Use of spatially distributed water table observations to constrain uncertainties in a rainfall-runoff model , *Advance in Water Resources* , Vol 22(4): 305-317.

Lindström G., B. Johanson, M. Person, M. Gardelin, S. Bergström, (1997). Development and test of the distributed HBV-96 hydrological model. *Journal of hydrology* 201 -272-288.

Noilhan, J. and J.-F. Mahfouf, 1996: The ISBA land surface parameterization scheme. *Global and Plan. Change*, 13, 145-159. http://www.cnrm.meteo.fr/isbadoc/pubs.html

Manabe, S. (1969) Climate and the ocean circulation, 1. The atmospheric circulation and the hydrology of the earth's surface. *Mon. Weath. Rev.* 97, 73Sept.774.

Milly PCD. (1993). An analytic solution of the stochastic storage problem applicable to soil water. *Water Resour. Res* . 29 (11) : 3755-3758.

Milly PCD. (1994). Climate, soil water storage, and the average annual water balance. *Water Resour. Res* . 30 (7) : 2143-2156.

Milly, P. C. D. (2001). A minimalist probabilistic description of root zone soil water, *Water Resour. Res* .., 37(3), 457-464.

Milly P. C. D and K. A. Dunne (2002). Macroscale water fluxes, 2, Water and energy supply control of their interannual variability. *Water Resour. Res* .38(10), 1206, doi:10.1029/2001 WR000760.

Montaldo Nicola Montaldo1 and John D. Albertson. Marco Mancini (2001). Robust simulation of root zone soil moisture with assimilation of surface soil moisture data. *Water Resour. Res* .37, 12, 2889-2900

Monteith, J.L., (1965). Evaporation and environment. Symp. Soc. Exp. Biol. XIX, 205-234. In: Evaporation. Selection, Introduction and commentary by J. H. C. Gash and W.J. Shuttleworth. Benchmark Papers in Hydrology. ISSN 1993-4572. IAHS Press 2007.

Nachabe, M. H. (1998), Refining the definition of field capacity in the literature, *J. Irrig. Drain. Eng.*, 124(4), 230- 232.

Nasta P., Kamai T., Chirico G. B., Hopmans J. W., Romano N., (2009). Scaling soil water retention functions using particle-size distribution. *Journal of Hydrology* 374 . 223

Olioso A., Braud I., Chanzy A., Courault D., Demarty J., Kergoat L. , Lewan E., Ottlé C., Prévot L., Zhao W.G., Calvet J. C., Cayrol P., Jongschaap R., Moulin S., Noilhan J., Wigneron J.P. (2002). SVAT modeling over the Alpilles-ReSeDA experiment: comparing SVAT models over wheat fields *Agronomie* 22 (2002) 651-668

Perrin Ch., Michel C., Andréassian V. (2003) .Improvement of a parsimonious model for streamflow simulation. *Journal of Hydrology* 279: 275-289

Perrochet P. (1987). Water uptake by plant roots − A simulation model, I. Conceptual model. *Journal of Hydrology*. Volume 95, Issues 1-2, 15 November 1987, Pages 55-61.

Philip, J.R., 1957. The theory of infiltration. *Soil Sci.* 1 (7), 83-85.

Pinol J., LLedo M.J. and Escarré A. (1991). Hydrological balance for two mediterranan forested catchments (Prades, Northeast Spain). *Hydrol. Sc. J.* 36(2)/4.

Raats P.A.C. (2001), Developments in soil–water physics since the mid 1960s. *Geoderma* 100. 355-387

Rawls W., Brakensiek D. et Saxton K. (1982) Estimation of soil water properties. *Trans. Am. Soc. Agric. Engrs* 25, 51-66

Rawls, W. J., and D. L. Brakensiek (1989), Estimation of soil water retention and hydraulic properties, in Unsaturated Flow in Hydrologic Modeling: Theory and Practice, edited by H. J. Morel-Seytoux, pp. 275– 300, Springer, New York.

Richards, L. A. (1931), Capillary conduction of liquids through porous medium, *Physics*, 1, 318– 333.

Rodriguez-Iturbe I, Proporato A, Ridolfi L, Isham V, Cox D. (1999). Probabilistic modeling of water balance at a point: the role of climate, soil and vegetation, *Proc. R. Soc. London Ser. A*, 455. 3789-3805.

Rodriguez-Iturbe I. (2000) Ecohydrology: A hydrologic perspective of climate-soil-vegetation dynamics. *Water Resour. Res* . 36, 1, 3-9

Saxton K.E., Rawls W.J., Romberger J.S., Papendick R.I. (1986) Estimating generalized soil-water characteristics from texture. Soil Sci. Soc. Am. J. 50, 1031-1036.

Seibert J., Multi-criteria calibration of a conceptual runoff model using a genetic algorithm, Hydrol. Earth Syst. Sci., 4, 215– 224, 2000.

Shiklomavov I. A. (1989) Climate and water resources. *Hydrological Sciences - Journal - des Sciences Hydrologiques*, 34, 5, 10/1989

Soil Survey Division Staff (SSDS). (1998). Keys to Soil Taxonomy, eighth ed. US Department of Agriculture, Washington DC, USDA-NRCS.234

Sutcliffe J. V. (2004). Hydrology: a question of balance. *IAHS special publication 7.* , IAHS Press, 2004.

Teshima, J., Hirayama, Y., Kobayashi, T., Cho, H. (2006) Estimating evapotranspiration from a small area on a grass–covered slope using the BBH model of soil hydrology. *J. Agric. Meteorol.* , 62 (2), 65-74

Van Genuchten M.T. (1980). A closed-form equation for predicting the hydraulic conductivity of unsaturated soils, Soil Soc. Am. J. 44 892–898

Villalobos F.J., Orgaz F., Testi L., Fereres E. (2000). Measurement and modeling of evapotranspiration of olive (Olea europaea L.) orchards. *European Journal of Agronomy* 13 (2000) 155–163

Vrugt, J. A., Schoups, G., Hopmans, J. W., Young, C., Wallender, W. W., Harter, T., Bouten, W. (2004) Inverse modeling of large-scale spatially distributed vadose zone properties using global optimization. *Water Resour. Res* .40, W06503, doi :10.1029/2003WR002706

Wagener T., Boyle D. P., M. J. Lees, H. S. Wheater, H. V. Gupta and S. Sorooshian (2001). A framework for development and application of hydrological models; *Hydrology and Earth System Sciences*, 5(1), 13–26.

Wagener, T.* N. McIntyre, M. J. Lees, H. S. Wheater1 and H. V. Gupta. (2003). Towards reduced uncertainty in conceptual rainfall-runoff modelling: Dynamic identifiability analysis. *Hydrol. Process.* 17, 455–476.

Wagener T., M. Sivapalan, P. Troch and R. Woods (2007). Catchment classification and hydrologic similarity. *Geography compass*: 901-931

Youngs E. G. (1988) . Soil physics and hydrology. *Journ. of Hydrol.* 100, 411-451.

Zhan, Ch.-S., Xia, J., Chen, Z., Zuo, Q.-T. (2008) An integrated hydrological and meteorological approach for the simulation of terrestrial evapotranspiration. *Hydrological Sciences Journal.*

Zhao, R. J., and X. R. Liu (1995), The Xinjiang model, in Computer Models of Watershed Hydrology, edited by V. P. Singh, chap. 7, *Water Resour. Publ.*, Highlands Ranch, Colo.

9

The Role of Evapotranspiration in the Framework of Water Resource Management and Planning Under Shortage Conditions

Giuseppe Mendicino and Alfonso Senatore

Department of Soil Conservation, University of Calabria, Arcavacata di Rende (CS)
Italy

1. Introduction

The increased availability of observed data and of advanced techniques for the analysis of meteo-hydrological information allows an even more detailed description of the evolution of global climate. The results showed by the Fourth Assessment Report (FAR) of the International Panel on Climate Change (IPCC, 2007) about the changes that, starting from 1950, are affecting the atmosphere, the cryosphere and the oceans, confirm global warming. The global average surface temperature has increased in the last 100 years by 0.74°C ± 0.18°C, accelerating in the last 50 years (0.13°C ± 0.03°C per decade), especially over land (about 0.27 °C per decade) and at higher northern latitudes. As a consequence, the higher available energy on the surface has speeded up the hydrological cycle. The concentration of the water vapor in the troposphere has increased (1.2 ± 0.3% per decade from 1988 to 2004), while long-period precipitation trends (both positive and negative) in many regions have been observed by analyzing time series from the year 1900 to the year 2005. Changes in temperature and precipitation regimes strongly affect the hydrological cycle. As an example, the increase in temperature has produced a substantial reduction in snow cover in several regions, mainly in spring, and a reduction in the areas covered by seasonal frozen ground (reduction of about 7% in the northern hemisphere over the latter half of the 20th century). Direct long-term measurements of all the main components of the hydrological cycle are not widely available: in order to assess soil moisture long-term changes, due to the lack of direct measurements the primary approach is to calculate Palmer Drought Severity Index, while long-term stream flow gauge records do not cover entirely and uniformly the world, and they present gaps and different record lengths. However, generally stream flow trends are positively correlated to precipitation, while a common effect of climate change is arising independently on precipitation trends: starting from the '70s a considerable increase of the frequency of extreme hydrological events (floods and droughts) has been observed. Also concerning actual evapotranspiration, direct measurements over global land areas are still very limited, but already the Third Assessment Report (TAR) reported that actual evapotranspiration increased during the second half of the 20th century over most dry regions of the USA and Russia, and, by means of observed precipitation, temperature, cloudiness-based surface solar radiation and a land surface model, Qian et al. (2006) found that global land evapotranspiration closely follows variations in land precipitation.

Following the FAR, it is extremely unlikely (<5% probability) that the global warming trend observed in the last half century, whose remarkable characteristics in the history of the Earth seem to be confirmed even by paleo-climatic studies, could be explained without considering external forcings, and is very likely (>90%) that the production of greenhouse gases is the main cause of the observed increase in temperature.

Human activities negatively impact on water resource availability, not only contributing to the water cycle changes on a global scale, but also in a more direct way, through the pollution of water courses and aquifers. This pollution is specifically generated by the over-exploitation of the soil and chemical contaminants due to agriculture and forestry, by urban waste, transportation and building, and by the over-exploitation of the coastal aquifers, which generates saline water intrusion.

Many of the problems connected to water shortage and to bad water quality are due to not efficient or even inexistent water resources planning and management. Recently, most advanced planning studies have adopted tools for integrated water resources management. Specifically, by now among planners the idea is diffused that a reactive approach, based on the implementation of actions after a drought event has occurred and is perceived, is not adequate and a proactive approach is needed (Yevjevich et al., 1983; Rossi, 2003), based on the development of plans allowing the identification of long- and short-term actions to face drought, and the implementation of such plans, on the basis of timely information provided by a drought monitoring system.

Different measures can be used to cope with water resource crises due to drought. Rossi et al. (2007) show several classifications of these measures: first, the one suggested by Yevjevich et al. (1978) that distinguishes among measures aimed at increasing water supply, reducing demand and minimizing impacts; next, considering the one differentiating reactive and proactive measures (Yevjevich et al., 1983); and finally, the one between long- and short-term measures. The Water Scarcity Drafting Group (2006) disseminated a document specifying a series of mitigation measures that can be adopted in the EU countries. Pereira (2007), starting from a conceptual distinction between water conservation (referred to the measures for the conservation and preservation of water resource) and water saving (referred to the measures aimed at limiting and/or controlling water demand), points out a set of actions that can be adopted in agriculture to reduce the impacts of drought resulting economically, socially and environmentally more competitive than the "classical" proposal of realizing artificial reservoirs, the latter being an alternative preferred in even fewer cases in the countries where water resource planning is more advanced (e.g. Cowie et al., 2002). Finally, the European Commission in the Communication "Addressing the challenge of water scarcity and droughts in the European Union", adopted on July 18, 2007 (COM, 2007), while stating the necessity of progressing towards the full implementation of the Water Framework Directive 2000/60/EC (WFD), underlines the huge potential for water saving across Europe, where people continue to waste at least 20% of water due to inefficiency, indeed leakages greater than 50% have been recorded in the irrigation networks. A report connected to the EU Communication (Dworak et al., 2007) estimates a potential water saving in the EU of about 40%. Regarding the strategic paths for future interventions, the enhancement of drought risk management should be achieved also through: developing drought risk management plans; developing an observatory (an European Drought

Observatory is now available at http://edo.jrc.ec.europa.eu) and an early drought warning system; further optimizing the use of the EU Solidarity Fund and European Mechanism for Civil Protection; fostering water efficient technologies and practices; fostering the emergence of a water-saving culture in Europe.

In this framework, evapotranspiration assessment is of outstanding importance both for planning and monitoring purposes. Its magnitude (mainly referring to potential evapotranspiration) is comparable to the main forcing of the water balance, i.e. precipitation, and for this reason several climatic classifications are based on comparisons between these two quantities, with the aim of determining specific climate conditions for different areas (e.g. Rivas-Martinez, 1995). Furthermore, evapotranspiration is the only component of the water balance with a central role also in the energy and carbon balance, since it directly accounts for hydrological, agricultural and ecological effects of drought events. Specifically, in agriculture evapotranspiration can be closely related to water demand. This means that the role of evapotranspiration, and losses due to evapotranspiration in agriculture (which are foreseeable to a certain extent) can be handled in a way allowing to assure the best conditions for agricultural needs, if water resources management is correctly planned and implemented. Hence, in this chapter evapotranspiration assessment/water demand fulfillment will be considered within the wider framework of water resources management and planning, both for a correct evaluation of the water balance (considering both the hydrological balance and the differences between water requirements and availability), and for determining incoming drought events through appropriate indices (drought monitoring). The issue of reducing water requirements, meaning loss reductions and/or evapotranspiration reductions (mainly in agriculture) will only be touched on, while dealing with methods and tools for water resource management under shortage conditions.

In the next sections, after an analysis of the available water resource and water demand in a southern Italian region (Calabria), the chapter highlights some weaknesses of the regional water system in rainfall deficit conditions, drafting the main strategies of intervention to be adopted to face the different aspects of drought. Then, some guidelines for the proactive management of drought in agriculture are proposed and specifically, by means of a case-study related to one of the most important agricultural areas in southern Italy (the Sibari Plain), the development of the three most important operational management tools is shown, i.e. the Strategic Plan for long-term interventions, the Management Plan for short-term interventions and the Contingency Plan for emergency conditions. Drought indices are important tools for correctly drafting these plans: a specific section will provide some insight about them. Finally, some climatologic and hydrologic scenarios over a specific basin are hypothesized, with the aim of assessing water resource availability in the second half of the present century and of verifying whether the intense and prolonged drought periods currently affecting the Calabria region will become ordinary situations in the near future.

2. Natural water resource

Since no useful information is available for an estimate of the direct runoff volume on the whole region, natural water resource was determined using a distributed monthly water balance model described by Mendicino & Versace (2007) and Mendicino et al. (2008a), which extends the approach proposed by Thornthwaite & Mather (1955) and simulates soil moisture variations, evapotranspiration, and runoff on a 5 km regular grid (Fig. 1) using data sets that include climatic drivers, vegetation, and soil properties. This model does not consider the horizontal motion of water on the land surface, or in the soil (hence no flow routing algorithms are required), and it is based on a simplified mass balance:

$$\Delta W = P + SM - SA - ET - Q \tag{1}$$

where ΔW is the change in soil moisture storage, P the precipitation, SM the snow melt, SA the snow accumulation, ET the actual evapotranspiration, and Q is the runoff (all the quantities are evaluated in mm month^{-1}). In the model, potential evapotranspiration PET is estimated through the Priestley-Taylor method (Priestley & Taylor, 1972), requiring only temperature, air pressure and net radiation data, overcoming the lack of observed wind speed and air humidity data in the analyzed area before the year 2000. In the case of net radiation, monthly values were obtained starting from a modified version of the model originally suggested by Moore et al. (1993). Actual evapotranspiration ET is calculated starting from PET and considering the Accumulated Potential Water Loss (APWL), such as suggested by Thornthwaite & Mather (1955), which represents the total amount of unsatisfied potential evapotranspiration to which the soil has been subjected.

Because of the significant reforestation campaigns carried out in Calabria after the Second World War, whose results were evident already at the end of 1950s, the starting period for the analysis was assumed to be 1957. The assumption of constant soil use (derived by the Corine Land Cover 2000 project) is justified by the coarse resolution of the model (5 km grid). The model schematized in figure 1 was improved also considering: i) that a portion of the rainfall is directly transformed into "instantaneous" runoff (depending on the ratio between actual soil moisture and soil water holding capacity WHC, in its turn derived by combining soil use with a detailed soil texture map of Calabria); ii) an additional very simple snow module, which partitions snow and rain precipitation and regulates snow melt just referring to the current monthly temperature in the cell; iii) that the hydraulic subsoil characteristics are simulated with reservoirs whose rates of depletion vary with the predominating geo-lithological characteristics in the single cells of the model (Mendicino et

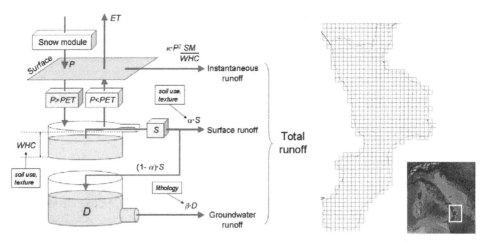

Fig. 1. Schematization of the water balance model and overlay of the 5 km regular grid in the analyzed region.

al., 2005). The different characteristics of subsoil leaded to the subdivision of the region into three categories: I) areas with a high capability of producing perennial flow (rocks with high permeability not in the plain); II) areas with mean capability of producing perennial flow (rocks with mean permeability); III) areas with low capability of producing perennial flow (rocks with low permeability or with high permeability in the plain).

The monthly water balance model was validated considering about 2900 monthly runoff values observed in 14 Calabrian catchments during the period 1955-2006 (Fig. 2). Figure 2 also shows the quite satisfactory performance of the model that, besides reproducing the monthly average behaviour of each considered catchment, provided values of the slopes of the regression curves obtained comparing observed and simulated runoff values varying from a minimum of 0.791 (Alli Orso) to a maximum of 1.135 (Esaro La Musica), while the correlation coefficients r varied from 0.447 (Coscile Camerata) to 0.939 (Corace Grascio).

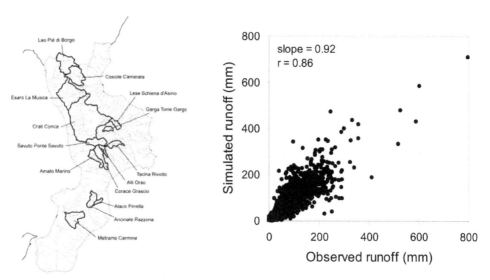

Fig. 2. Spatial distribution of the gauged catchments and comparison between all observed and simulated runoff during the period 1960-2006.

The monthly water balance model was applied on the whole territory of Calabria for the period 1960-2006 on a 5 km regular grid, where each cell was independent from the others, determining the main components of the hydrological balance in the whole region: precipitation, actual evapotranspiration, soil moisture storage, groundwater volume and instantaneous, surface and subsurface runoff. In several areas of the region a negative trend was observed for many of these variables. Specifically, while the potential evapotranspiration trend was strongly related to increasing temperature, actual evapotranspiration was affected also by changes (reduction) in precipitation. Considering the whole region, the average annual actual evapotranspiration estimated in the analyzed period is 581 mm, equal to about 57.8% of the average cumulated annual rainfall (potential evapotranspiration is about 110%). Figure 3 (left side) shows the average monthly values in the whole region for actual and potential evapotranspiration. The months where a significant difference can be observed are the months from May to September. In these months (the less rainy and warmest ones), evaporation of soil moisture accumulated in wintertime exceeds rainfall, requiring irrigation in most of the agricultural areas. Figure 3 (right side) also shows the trend of cumulated annual actual evapotranspiration. The decrease in time of this quantity due to rainfall reduction is partly balanced by the increasing temperatures, hence the negative trend is not significant. It is noteworthy that peaks and troughs are generally dependent on rainy (e.g. 2005) or not rainy (e.g. 2001) years, even though rainfall distribution during the single year also affects the evapotranspirative phenomenon. The correlation coefficient between cumulated annual actual evapotranspiration and precipitation in the period 1960-2006 was 0.638.

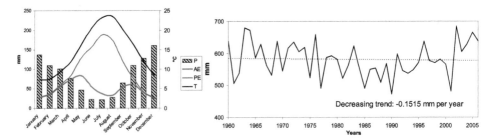

Fig. 3. Left: average monthly values in Calabria of actual (AE) and potential evapotranspiration (PE), precipitation (P) and temperature (T) during the period 1960-2006. Right: trend of cumulated annual actual evapotranspiration.

3. Water demand and availability

The water balance between available water resource and water demand is the starting point for a correct water management. One of the main problems occurring in this phase is the general lack of observed data, obliging to synthetic estimates of water availability and several levels of approximation in the assessment of water needs, mainly for irrigation and for determining the management rules of the reservoirs.

In this context, the water balance on the Calabrian region was carried out considering also withdrawals from springs, streams, reservoirs and wells for irrigation and for potable uses, adopting two sequential simulation models. The former is a modified version of the distributed hydrological model, where the natural water balance is integrated with the withdrawal for irrigation and potable uses, producing (output variable) a residual availability. This water availability is used in a second GIS-based model considering the effects of diversions and reservoirs.

In the first model, inside a single 5 km squared cell can co-exist both wells and springs used to feed small irrigation systems or few users, located in the same cell, and wells and springs used for water mains collecting water outside the cell. If both the points where the water is withdrawn and used are inside the same cell (this happens only for wells for irrigation purposes), the schematization shown in figure 4a is adopted, hypothesizing that inside the cell a known volume is transferred monthly from the subsoil reservoir to the surface as an "added" precipitation (owing to the irrigation). This volume has to be summed to the meteorological precipitation and is subjected to the cycle simulated by the water balance, increasing the soil moisture and actual evapotranspiration and eventually feeding the aquifer from which it has been withdrawn. Instead, if the cell where the water is withdrawn does not coincide with the cell where it is used (that is only the case of regional water mains) then the schematization shown in figure 4b is adopted. The source cell is subjected to a reduction of the volume of the subsoil reservoir, while the water is hypothesized to reach directly the water stream in the destination cell, feeding the surface runoff with a restitution coefficient equal to 0.7.

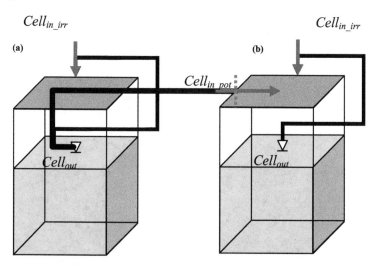

Fig. 4. Schematization of the modified water balance model considering withdrawals for irrigation and potable uses.

Summarizing, the proposed model allows that every month for each cell a volume $cell_{out}$ can be extracted from the subsoil reservoir, which is equal to the withdrawals for irrigation and potable purposes, that a volume $cell_{in\_irr}$ can be added like a supplementary precipitation representing the water derived from the same cell and used for irrigation, and finally, that a volume $cell_{in\_pot}$ can be added like a supplementary surface runoff accounting for the water come in the cell to satisfy the potable uses. All the data related to potable and irrigation withdrawals were derived from several official sources, even if sometimes incomplete, and were aggregated at the resolution of the water balance model. Figure 5 shows the distribution of the regional water mains and of the local water distribution systems.

The modified natural water balance is the input of the commercial GIS-based model Mike Basin (DHI Software), accounting for the effects of diversions and reservoirs aimed at satisfying irrigation, hydro-power, civil and industrial requirements (Fig. 6). The lack of actual information about the management rules of reservoirs led to hypothesize several working schemes for the definition of the optimal water balance. Finally, for all the analyzed reservoirs the minimum flow requirements were considered following two different approaches: the former proposed by the Regional Basin Authority (very conservative, especially for some typical Calabrian rivers, called *fiumare*, characterized by no flow conditions for a relevant part of the year) and; the latter based on the $Q_{7,10}$ flow, i.e. the lowest 7 consecutive-day average flow characterized by a 10 years time period.

In the case of the irrigation demand (i.e. water requirements for balancing evapotranspiration losses), a detailed analysis was carried out on each irrigation district

(Fig. 7) during the irrigation season April – September. Specifically, the assessment of the effective water consumption was determined by considering different seasonal (spring, summer, autumn) soil use spatial distributions (e.g. in Table 1). For each soil use the seasonal irrigation demand (m³/ha) of the crops (Table 2) was achieved. The same was split monthly taking into account that the highest request is obtained during the trimester June – August (Table 3). An adequately detailed knowledge of the irrigation network allowed the correct estimate of the possible uptake of volumes to/from other cells. It is noteworthy to highlight that all the information related to soil use and water requirements were aggregated at the resolution of the model, i.e. 5×5 km², for the whole region.

In the proposed analysis the quite small volumes related to industrial areas were neglected.

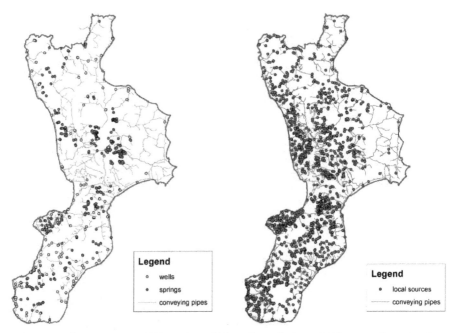

Fig. 5. Left: regional water mains (479 springs, 281 wells and about 2000 conveying pipes). Right: local water distribution systems (over 1200 springs and wells).

Water balance results showed that, for average conditions, the residual annual water availability is great, even if some weaknesses arise. Among these, the strong differences in the seasonal precipitation, which is mainly concentrated in the wet winter period (80-90%), require an accurate management of the volumes stored in natural and artificial reservoirs for facing the hot and dry Mediterranean summer. Furthermore, the decrepitude of several conveying pipes has to be considered with remarkable water losses, and the negative precipitation trend due to climate change that seems to be relevant in Calabria (a preliminary analysis about future climate scenarios in Calabria is shown in the 6th section).

The weaknesses pointed out in normal conditions suggested water resources availability should be analyzed when drought conditions occur. Specifically, through the use of the Standardized Precipitation Index (SPI, McKee et al., 1993) intensity and duration of droughts were determined on the whole Calabrian region.

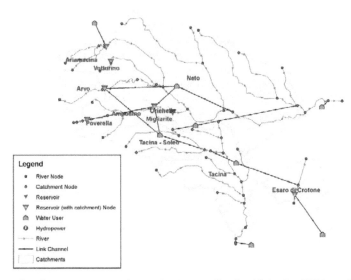

Fig. 6. Example of water system schematization realized within the GIS-based model Mike Basin.

Crops		Spring	Summer	Autumn
Code	Description	Soil use (ha)		
2121	Spring-summer herbaceous crops		14.115	
2122	Summer-autumn/spring horticultural crops			14.115
2123	Spring-summer horticultural crops			
2211	Irrigated vineyards	4.536	4.536	4.536
2221	Irrigated orchards	328.791	328.791	328.791
2231	Irrigated olive groves	95.788	95.788	95.788

Table 1. Seasonal soil use for a generic irrigation district.

For each of the most significant Calabrian basins, and for each month of the period 1960-2006, a mean SPI areal value was calculated for different time scales (1-, 3-, 6-, 12-, 24- and 48-months), with the aim of highlighting the longest and most intense drought periods (Fig. 8). Drought indices are essential at all levels of the planning process. The reader is referred to section 5 for a brief review of the most diffused ones.

Usually, the beginning of a drought period can be defined when SPI values are lower than -1.0, and its end when the values come back positive. Nevertheless, based on a historical analysis of the official declarations of "natural disaster" in Calabria due to drought, even a 12-month SPI value equal to -0.7 was observed to be adequate as a drought threshold. Hence, when a generic month presented a 12-month SPI value lower than -0.7, it was considered a drought month, and the correspondent total runoff simulated with the water balance model was taken into account. The aggregation, from January to December, of the average runoff estimated during the drought months leaded to the definition of a so-called "scarce year" whose runoff values, even if statistically less probable than the ones of the single months, pointed out the possibility of extremely critical situations in Calabria, with a reduction of total runoff up to 43%. This analysis introduces issues related both to the management of water shortage and to the mitigation of drought through the use of restrictive measures. The development and implementation of strategic and emergency plans are primary tools to face the different aspects of drought phenomenon, as it is shown in the next paragraph.

Fig. 7. Calabrian irrigation districts and network systems.

Code	Description	Irrigation demand (m³/ha)
2121	Spring-summer herbaceous crops	7000
2122	Summer-autumn/spring horticultural crops	7600
2123	Spring-summer horticultural crops	5000
2125	Greenhouse crops	9000
213	Rice fields	15000
2211	Irrigated vineyards	3500
2221	Irrigated orchards	5000
2231	Irrigated olive groves	3000

Table 2. Seasonal irrigation demand (m³/ha) of the crops.

Code	Description	A	M	J	J	A	S	TOT
2121	Spring-summer herbaceous crops	0	858	1497	2337	1445	863	7000
2122	Summer-autumn/spring horticultural crops	163	745	1719	2450	1663	861	7600
2123	Spring-summer horticultural crops	164	751	1733	2352	0	0	5000
2125	Greenhouse crops	193	882	2035	2901	1969	1020	9000
213	Rice fields	3780	3240	3240	3240	1500	0	15000
2211	Irrigated vineyards	0	0	1063	1411	1026	0	3500
2221	Irrigated orchards	0	0	1349	2125	1168	458	5000
2231	Irrigated olive groves	0	0	894	1249	857	0	3000

Table 3. Monthly irrigation demand (m³/ha) of the crops.

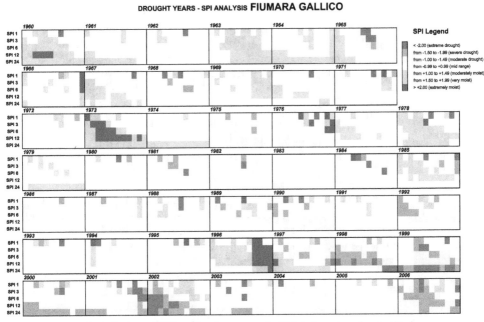

Fig. 8. Temporal evolution of SPI values in a generic Calabrian river basin. Red squares correspond to drier periods.

4. Water resource management under shortage conditions

In its 2007 Communication (COM, 2007) the European Commission stated that the challenge of water scarcity and droughts needs to be addressed both as an essential environmental issue and as a precondition for sustainable economic growth in Europe, and highlighted the necessity of progressing towards full implementation of the EU Water Framework Directive (WFD) 2000/60. The WFD is the EU's flagship Directive on water policy, explicitly defining long-term planning as the main tool for ensuring good status of water resources. Nevertheless, it does not indicate criteria and actions to face risk of drought, delegating National Legislations to concretely realize its framework (after a series of yearly follow-up reports, a policy review is foreseen for 2012 at the EU level).

In Italy the EU WFD was taken into account with the Legislative Decree 152/2006 on environmental protection. Though this act is quite recent, it seems to be far from being adequate to actually cope with drought, mainly because it does not stress the necessity of passing from a reactive to a proactive approach, based on preparedness and mitigation actions planned in advance with the contribution of all the involved stakeholders, ready to be implemented when drought phenomena occur.

Within a comprehensive drought management planning process, Rossi et al. (2007) proposed the identification of three main tools: Strategic Water Shortage Preparedness Plan, Water Supply System Management Plan and Drought Contingency Plan. Following, an example of application of the proposed guidelines is shown for the planning of the best mix of measures needed for coping with drought phenomena on one of the most important agricultural areas in southern Italy, the Low Esaro and Sibari Plain (Mendicino et al., 2008b). It is noteworthy that in the proposed example (water shortage planning in the agricultural sector) water demand is strictly correlated to the amount of water needed from crops for facing lack of precipitation and high potential evapotranspiration during summer (see Table 3). Hence, in this case the planning process is triggered by the need of coping with the high water loss due to evapotranspiration in a particularly dry period of the year. As it is explained in the next sections, this objective can be reached by means of demand reduction, water supply increase or impacts minimization measures, and considering long-, medium- and short-term actions.

4.1 Methods and tools

The Agricultural Strategic Water Shortage Preparedness Plan (ASP) is aimed at obtaining the reduction of drought vulnerability in the analyzed area through the implementation in normal conditions of long term mitigation measures, consisting in a series of structural and non-structural actions applied in the water supply system. Usually, structural measures are economically expensive and require the use of many human resources. However, their effects are easier to be foreseen than the effects produced by the non-structural mitigation actions, in their turn usually more accepted by all the stakeholders. The long term mitigation measures are specifically indicated in the systems characterized by a low level of reliability and are oriented at improving the water balance in the analyzed system. These actions not only enhance the reliability of the system through fulfilling water requirements, but also reduce its vulnerability with respect to future drought events, fulfilling three main objectives: i) water demand reduction; ii) water supply increase and improvement of the efficiency of the system; iii) minimization of the impacts. Within the actions reducing water demand, some are directly aimed at reducing evapotranspiration by adopting appropriate agronomic techniques, such as e.g. irrigating during non windy periods for minimizing wind drift losses, or early defoliation to reduce crop transpiration surface (for a deeper description, the reader is referred to Pereira, 2007). In table 4 the long term measures that can be potentially adopted in agriculture are listed, subdivided considering their main objectives.

The Role of Evapotranspiration in the
Framework of Water Resource Management and Planning Under Shortage Conditions
191

Category	Long-term actions
Demand reduction	Economic incentives for water saving and sanctions for wastes
	Agronomic techniques and irrigation systems for reducing water consumption (e.g. Pereira, 2007)
	Dry crops in place of irrigated crops
Water supply increase and improvement of the efficiency of the system	Conveyance networks for bi-directional exchanges
	Reuse of treated wastewater
	Inter-basin and within-basin water transfers
	Construction of new reservoirs or increase of storage volume of existing reservoirs
	Use of aquifers as groundwater reservoirs
	Non conventional sources (particularly desalination of brackish or saline waters)
	Control of seepage and evaporation losses
	Elimination of the possible risks of pollution of the sources
	Modernization and restructuring of the irrigation network
Impacts minimization	Reallocation of water resources based on water quality requirements
	Development of early warning systems
	Implementation of Agricultural Management Plans and Contingency Plans
	Insurance programs
	Education activities for improving drought preparedness and/or permanent water saving

Table 4. Main long term drought mitigation measures in agriculture (adapted from Rossi et al., 2007, and Georgia Dept. Of Natural Resources, 2003).

Since the ASP has to be drawn up choosing among several combinations of long-term mitigation measures, a suitable evaluation procedure has to be adopted. A multi-criteria technique could provide an as objective as possible comparison among different alternatives, according to a series of economic, environmental and social criteria, and taking into account the point of view of all the stakeholders. The tool adopted in this study for multi-criteria analysis is the software NAIADE (Munda, 1995).

The ASP should be prepared by the Basin or Hydrographic District Authorities, which are the bodies responsible for planning, and corresponds to the Drought Management Plan included into the River Basin Management Plan provided in the WFD.

Once the long-term mitigation measures are defined, an Agricultural Water Supply System Management Plan (AMP) has to be developed with the aim of: defining the best mix of long and short-term measures to avoid the beginning of a real water emergency; estimating the costs and the financing sources for the chosen mitigation measures, and; fostering the stakeholder participation and exchanges. It is prepared by the authority responsible for agricultural water management (i.e. the Land Reclamation Consortium), and the operative measures defined have to be adopted according to the values of early warning indicators, showing Normal, Pre-Alert or Alert conditions. The threshold values of the indicators can be chosen through an objective function or, if several aspects have to be accounted for, through a multi-criteria analysis.

In table 5 the short term measures that can be potentially adopted in agriculture are shown, subdivided on the basis of their principal objectives. With respect to the long-term mitigation measures, in this case the actions in the "demand reduction" category implicitly accept a certain percentage of water stress for the crops, because they are only aimed to reduce water consumption, without taking into account crop conditions. On the contrary, former long-term mitigation measures suggested some structural actions (i.e. actions to be adopted always) aimed at limiting some additional evapotranspiration due, e.g., to not correct irrigation practices, and that could be avoided without consequences for the crop. In brief, adopting the AMP evapotranspiration losses could be not completely compensated, and the farmers should be supported in assessing how to minimize water stress effects adopting even more specific agronomic techniques.

Category	Short-term actions
Demand reduction	Public information campaign for water saving
	Restriction of irrigation of annual crops
	Pricing (discourage excessive water use)
	Mandatory rationing
Water supply increase	Improvement of existing water systems efficiency (leak detection programs, new operating rules, etc.)
	Use of emergency sources (additional sources of low quality and/or high exploitation cost)
	Over exploitation of aquifers (use of strategic reserves)
	Increased diversion by relaxing ecological or recreational use constraints
Impacts minimization	Temporary reallocation of water resources
	Public aids to compensate income losses
	Tax reduction or delay of payment deadline
	Public aids for crops insurance

Table 5. Main short term mitigation measures in agriculture (adapted from Rossi et al., 2007).

If a particularly severe drought occurs, and the indicators signal Alarm conditions, the Agricultural Drought Contingency Plan (ACP) has to be adopted, defining the most appropriate short-term measures to reduce the impact of emergency situations. In this case the efforts are turned to protect the essential activities of the agricultural system, and the threshold values of the indicators have to be chosen taking into account this objective, preferably using a probabilistic approach, that allows the decision-makers to evaluate the effective risk of having water deficit for different scenarios. The ACP should be prepared by the Basin or Hydrographic District Authorities, with the collaboration of the Civil Protection.

Such as in the AMP, also in the ACP the assessment of crop losses can be made through production functions. In the case of extreme and particularly prolonged drought also the damage to perennial crops, the excessive decrease of the water tables of the aquifers, sea water intrusion, ecological damages to aquatic flora and fauna have to be considered. Some of this damage can be irreversible and can also influence crop production in the following years.

The Role of Evapotranspiration in the
Framework of Water Resource Management and Planning Under Shortage Conditions
193

4.2 Case study

The core of the analyzed water supply system is the Farneto Dam (Fig. 9), closing the Esaro Catchment (about 245.4 km²) in southern Italy. The dam is aimed at: (i) containing the ordinary floods and mitigating the extraordinary ones, according to the condition that the reservoir level is maintained almost empty from October to March; (ii) supplying water (about 30 hm³ from April to September) to the downstream agricultural area (about 85 km²), sited in the Low Esaro and Sibari Plain. At present about 63% of the irrigable area is based on open channel irrigation systems.

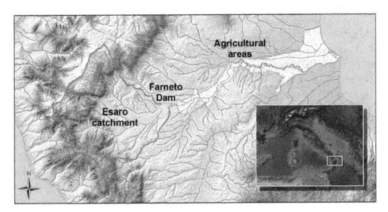

Fig. 9. Study area for the development of the planning process.

4.3 Applying the Agricultural Strategic Plan

Table 6 shows 13 selected alternatives (from A to M), obtained combining the following six long-term mitigation measures: 0) System in current configuration; 1) Modernization of the irrigation network for reducing water losses and evaporation (it has been calculated that the efficiency of the actual scenario is equal to 67%, while the efficiency of the "modernized" scenario will be 80%; Mendicino et al., 2008b); 2) Construction of farm ponds; 3) Construction of a new upstream dam; 4) Economic incentives and educational activities for water saving; 5) Allowing the dam to store a little volume during the winter (i.e. dam not empty in March).

	Alternatives												
Measure	A	B	C	D	E	F	G	H	I	J	K	L	M
0	X												
1		X				X	X			X			X
2			X								X		
3							X					X	X
4				X		X			X	X	X	X	X
5					X		X	X	X	X	X	X	X

Table 6. Long-term mitigation measures and alternatives.

The alternatives were compared within the DSS tool NAIADE according to 4 economic criteria (construction costs of infrastructures, operation and maintenance costs, crop yield

losses and amount of public aids needed), 2 environmental criteria (failures to meet
ecological requirements and reversibility of the alternatives) and 4 social criteria (system
vulnerability, temporal reliability, realization time of the infrastructures and employment
increase). Since the observed period is short in order to evaluate the criteria and is
characterized by few drought events, two monthly synthetic temperature and precipitation
series of 1000 years were generated as input of the water balance model providing the
corresponding runoff values.

Within the analysis carried out with NAIADE the final ranking of the alternatives comes
from the intersection of two separate rankings. The former Φ^+ is based on the "better" and
"much better" preference relations, hence it points out how an alternative is "better" than
the others. The latter Φ^- is based on the "worse" and "much worse" preference relations, and
indicates how an alternative is "worse" than the others.

The two rankings are different, since one alternative could result slightly better than the
others with respect to few criteria and at the same time could result worse with respect to
many criteria, or vice versa. In figure 10 the partial rankings and the final ranking are
shown. The most efficient alternative is the "J", where measures 1, 4 and 5 are considered
together. The alternative "M", mainly characterized by the construction of a new upstream
dam, is the best only in the Φ^+ ranking. A sensitivity analysis, carried out to assess the
robustness of the achieved solution, showed a substantial stability of the ranking, constantly
confirming alternative J as the optimal one. It is pointed out that alternative J is made up
also by measure 1), allowing a reduction of evaporation losses.

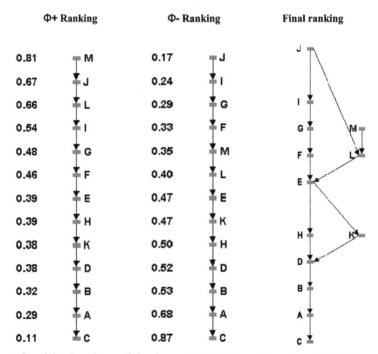

Fig. 10. Partial and final ranking of the drought mitigation alternatives in the Esaro River
Basin.

4.4 Applying the Agricultural Management Plan

The AMP is aimed at defining the indicators and the triggers for establishing the Normal, Pre-Alert and Alert conditions for the agricultural areas of the system. It has to take into account the guidelines provided by the ASP. In fact, it has to select the best combination among the optimal long-term mitigation measure previously determined (J) and the several short-term measures that can be adopted to manage water deficits on the analyzed area. Whereas the long-term measure J is adopted continuously, the short-term measures vary following the status of the system. Specifically, for this case study:

-   in Normal condition no short-term actions are taken;
-   when Pre-Alert condition occurs, then exploitation of the groundwater resources in the irrigated area till 1/3 of maximum estimated volume is considered;
-   when Alert condition occurs, then exploitation of the groundwater resources like in the Pre-Alert condition, the reduction of the release for minimum instream flow till 50% and the reduction of the release for irrigation (till 80% of the requirements) are taken into account. When alert condition occurs, the farmer is aware that the evapotranspiration losses cannot be completely compensated.

With the aim of determining the threshold values of the indices indicating the passage from one status to another, for every month from April to September a multicriteria analysis of the effects through NAIADE was carried out. The conflicting objectives to minimize are:

-   the vulnerability of the system (including the assessment of crop losses due to reduced irrigation, made through specific production functions);
-   groundwater withdrawals;
-   the failures to meet the minimum instream flow.

For each month, starting from April, an impact matrix was achieved where, on the basis of the criteria selected for the fulfillment of the objectives, the optimal combination of the thresholds triggering the Pre-Alert and Alert status was selected (Fig. 11). The selected index for the definition of the drought thresholds is the volume stored in the dam from May to

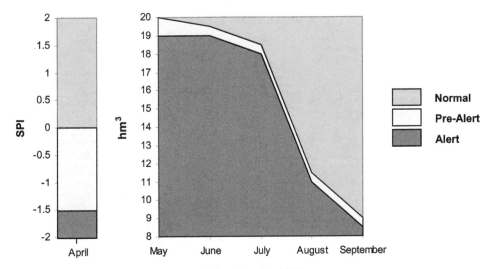

Fig. 11. Pre-Alert and Alert thresholds defined in the AMP.

September, while for the month of April a meteorological index was chosen, since owing to the rules adopted for dam management, at the end of March the dam level is not a significant index. For the month of April an analysis was carried out relating the yearly irrigation deficit to the 6 month-SPI calculated in March (considering in this way the first six months of the hydrologic year, from October to March). In the selection of the threshold values a rule was followed considering that, if the multicriteria analysis provides more optimal solutions, the one with the lowest irrigation deficit is selected.

4.5 Applying the Agricultural Contingency Plan

The first objective of the ACP is the definition of indices and their thresholds for univocally establishing the beginning of an emergency situation. Since the hydrologic analysis in April shows that the water demand is always less than the water availability in the Farneto del Principe Dam, and that every year the volume stored increases during this month, the thresholds are selected starting from May, choosing as an index, such as in the AMP, the volume stored in the dam. Furthermore, since using the 1000-year series of generated meteorological data the application of the two previous Plans determined a very high temporal reliability of the system (98.7%), it is not useful to evaluate the emergency thresholds considering the few residual years. Hence, the adopted approach was based on a probabilistic analysis of the system failures and deficit percentage of the demand.

Specifically, hypothesizing that all the short-term measures were already adopted, the 1000-year series of generated meteorological data, for every month and for different fixed initial volumes stored, were used to assess the probability of having failures in fulfilling demand either in the same month or in the subsequent irrigation period, and the deficit percentage with respect to demand. The results, allowing the decision-makers to evaluate the effective risk of having water deficit for a specific storage in a specific month, are shown (from May to August) in figure 12.

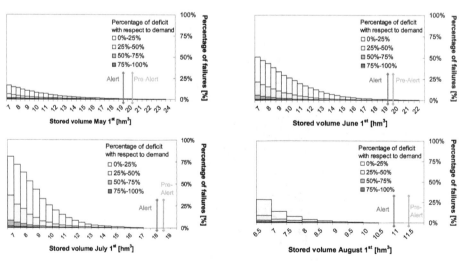

Fig. 12. Monthly risk of having failures and deficit percentage with respect to demand (from May to August).

5. Drought indices

Drought indices are tools necessary at all levels of the planning process: as it was shown in the previous sections, in the Strategic Plan they are used to identify the zones most exposed to drought risk in the analyzed areas, whereas in the Management Plan and in the Contingency Plan they are used to define trigger values for the activation of the measures for impact prevention or mitigation.

Most of the proposed methodologies for the characterization and the monitoring of drought phenomena are based on drought indices with the capability of synthetically summarizing drought conditions in a specific moment for a particular area. Nevertheless, drought is difficult to represent through a single index, hence frequently more indices or aggregate indices are used.

In rainfed agriculture meteorological indices are particularly suitable, because they give the opportunity of establishing a direct spatial correlation between the drought event and the agricultural production, allowing drought risk maps to be drawn.

Many authors provide lists describing the characteristics of the main drought indices (e.g. Ntale & Gan, 2003; Tsakiris et al., 2007a). Among them, the most widely used are the Palmer Drought Severity Index (PDSI; Palmer, 1965), the most "classical" drought index formulated to evaluate prolonged periods of both abnormally wet and abnormally dry weather conditions, and the Standardized Precipitation Index (SPI; McKee et al., 1993), a meteorological drought index based on the precipitation amount in a period of n months. Since SPI just needs precipitation data to be calculated, it has found widespread application. Guttman (1998) shows that the PDSI has a complex structure with an exceptionally long memory, while the SPI is an easily interpreted, simple moving average process. Hayes et al. (1999) describe the three main advantages in using SPI: the first and primary is its simplicity, the second is its variable time scale, and the third is its standardization. Nevertheless, the SPI is a meteorological index unable to take into account the effects of aquifers, soil, land use characteristics, crop growth and temperature anomalies, which influence agricultural and hydrological droughts.

Besides SPI, in the process of drought identification the MEDROPLAN Guidelines (Tsakiris et al., 2007a) suggest using also: the Reconnaissance Drought Index (RDI, Tsakiris et al., 2007b), also accounting for temperature anomalies (therefore for an eventual excessive evapotranspiration); deciles (Gibbs & Maher, 1967), used by the Australian Drought Watch System, which compare monthly observed precipitation values with the quantiles corresponding to the not exceeded frequencies of 10%, 20%,... 100% achieved from a long enough monthly precipitation series; the Surface Water Supply Index (SWSI, Shafer & Dezman, 1982), aggregating information about precipitation, runoff, volumes stored in the reservoirs and snowpack, and expressing drought conditions in a standardized way. Furthermore, owing to their diffusion, other two indices are recalled: the run method (Yevjevich, 1967), based on the comparison between the time series of the analyzed hydrological index and a representative threshold of "normal" conditions, and the Palmer Hydrological Drought Index (Karl, 1986), a modified version of the PDSI for real-time monitoring.

An interesting way to account for soil and land use effects (in some respects, the way followed by Palmer to calculate PDSI) is to derive the drought indices starting from hydrological modeling. These indices can be called "comprehensive" drought indices, because they allow a more comprehensive picture of the water cycle and its elements

(Niemeyer, 2008). A typical example of comprehensive drought index is the Groundwater Resource Index (GRI) derived by Mendicino et al. (2008a) using the monthly water balance model shown in figure 1. For each single element where the model was applied (5 km regular cell), the monthly values of groundwater detention (i.e. the storage D) were standardized (for almost all the cells and months the skewness test of normality showed that the series were normally distributed) through the following equation:

$$GRI_{y,m} = \frac{D_{y,m} - \mu_{y,m}}{\sigma_{y,m}} \qquad (2)$$

where $GRI_{y,m}$ and $D_{y,m}$ are respectively the values of the index and of the groundwater detention for the year y and the month m, while $\mu_{D,m}$ and $\sigma_{D,m}$ are respectively the mean and the standard deviation of groundwater detention values D simulated for the month m in a defined number of years (at least 30). This simple index, but based on several pieces of information provided by the water balance model, allows assessment of the deviation from the mean values of the available groundwater in a spatially-distributed way for the whole territory where the model is applied. Figure 13 shows the maps of the GRI distribution in northern Calabria for the months of April from 1979 to 2006. Examining the maps immediately the years with lower GRI values (the driest years, with brighter colors) are recognizable, as are the wettest years (darkest colors).

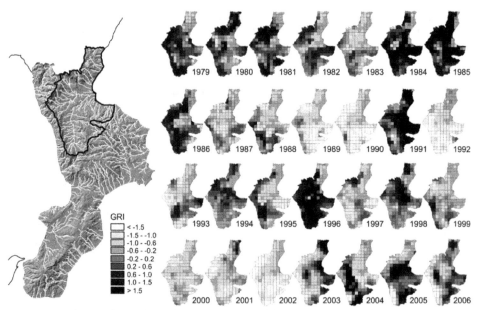

Fig. 13. Boundaries of the selected study area in Calabria and GRI distribution in north-eastern Calabria for the months of April from 1979 to 2006 (from Mendicino et al., 2008a).

Other comprehensive indices were developed by Narasimhan & Srinivasan (2005), who using the Soil and Water Assessment Tool (SWAT) model, derived two drought indices for agricultural drought monitoring, the Soil Moisture Deficit Index (SMDI) and the Evapotranspiration Deficit Index (ETDI), based respectively on weekly soil moisture and evapotranspiration (ET) deficit. Also Matera et al. (2007) derived a new agricultural drought index, called DTx, based on the daily transpiration deficit calculated by a water balance model.

In the few last years the possibility of using long data series coming from remote sensing has opened new and promising perspectives to satellite-derived drought indices, which have the advantage of being intrinsically spatially distributed. Anderson et al. (2007) provide a brief presentation of TIR-based drought indices, while a list of many NOAA-AVHRR images-derived drought indices is presented by Bayarjargal et al. (2006). Zhang et al. (2005) exploit the capabilities of the MODerate resolution Imaging Spectroradiometer (MODIS) for monitoring and forecasting crop production using a satellite-based Climate-Variability Impact Index.

Several remote sensing-derived drought indices depend on the ratio ET/PET, where ET is actual evapotranspiration and PET potential evapotranspiration (e.g. Crop Water Stress Index (CWSI), Jackson et al., 1981; Drought Severity Index (DSI), Su et al., 2003; Evaporative Drought Index (EDI), Anderson et al., 2007; Yao et al., 2010). While PET is generally calculated by means of ground based measurements, ET is easily estimated through "residual" methods (e.g. SEBAL, Bastiaanssen et al., 1998; and Bastiaanssen, 2000; SEBI, Menenti & Choudhury, 1993; S-SEBI, Roerink et al., 2000; SEBS, Su, 2002; TSEB, Norman et al., 1995; DisAlexi, Anderson et al., 1997; METRIC, Allen et al., 2007), where the evapotranspirative term is the residual term of the energy balance equation:

$$\lambda E = R_n - G - H \tag{3}$$

with R_n net radiation, G soil heat flux, H sensible heat flux and λE latent heat flux, from which ET is derived.

Even though at this stage very seldom they are used as operational tools, remote sensing-derived indices are potentially very useful because they intrinsically provide space-time variation of drought phenomena, and the ratio ET/PET can be reasonably related to soil water content. For instance, the relative evaporation Λ_r can be directly linked to the soil degree of saturation θ/θ_s (Su et al., 2003). As an example, figure 14 shows the space-time evolution of the DSI, derived from SEBS and MODIS images, during summer 2006 in Northern Calabria. DSI is equal to $1 - \lambda E / \lambda E_{wet}$ (where λE_{wet} is the latent heat flux estimated for the so-called "wet" pixel), hence higher DSI values indicate low actual evapotranspiration. A graph shown at the top of the figure provides information about precipitation in a micrometeorological station placed almost in the middle of the area (these data are only roughly representative, owing to the extension of the whole area). Figure 14 shows that the maps with the highest DSI values (e.g. July 20, but also September 4 and October 31), indicating drought stress conditions, are related to some of the most distant days from antecedent significant precipitation events.

To complete this brief review, a much-discussed issue is mentioned, i.e. the possibility of using the drought indices (especially SPI) to forecast stochastically the possible evolution of an ongoing drought (Cancelliere et al., 1996; Lohani et al., 1998; Bordi et al., 2005; Cancelliere et al., 2007). Several studies are also aimed at explaining and predicting possible drought conditions through the analysis of sea surface temperature (SST) and atmospheric circulation patterns (e.g. Wilby et al., 2004; Kim et al., 2006; Cook et al., 2007).

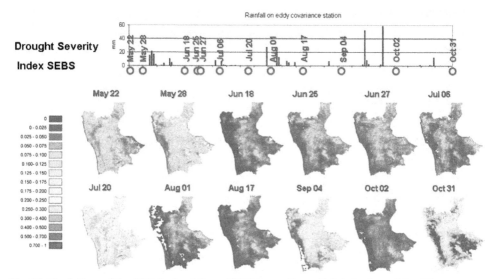

Fig. 14. Evolution of the DSI derived from SEBS in northern Calabria, from May 22 to October 31 2006. Top graph shows precipitation events on the representative micrometeorological station, placed approximately in the middle of the analyzed area.

However, when dealing with complex systems, where irrigated agriculture assumes a greater importance, one single index is often not able to capture the different features of drought and to take in account the effects of human activities (use of irrigation, water from reservoirs, wells, etc.) on the hydrological cycle. On the other hand, it is more practical to declare drought condition considering only one indicator. Thus, there is a growing interest in aggregating more indices. Keyantash & Dracup (2004) use an Aggregate Drought Index that considers all relevant variables of the hydrological cycle through Principal Component Analysis (but they do not include groundwater in the suite of variables); instead Steinemann & Cavalcanti (2006) use the probabilities of different indicators of drought and shortage, selecting the trigger levels on the basis of the most severe level of the indicator or the level of the majority of the indicators.

6. Future scenarios

The most critical scenarios discussed in the previous paragraphs could become "normal" circumstances if global climate change increases the prolonged and intense drought periods. At the end of the proposed analysis, it is useful to hypothesize some future climatic scenarios, with the aim of steering decision makers towards suitable water management policies, as it is suggested by the European Commission (COM, 2009).

The methodology usually followed to assess the hydrological consequences of climate change basically consists of a three-step process (Xu et al., 2005): (1) the development and use of general circulation models (GCMs) to provide future global climate scenarios under the effect of increasing greenhouse gases, (2) the development and use of downscaling techniques (both statistical methods and nested regional climate models, RCMs, which are being continuously improved) for "downscaling" the GCM output to the scales compatible with hydrological models, and (3) the development and use of hydrological models to simulate the effects of climate change on hydrological regimes at various scales. However, uncertainties within this framework have to be taken into account such as the internal variability of the climate system, model structure and parameterizations at different spatial and temporal scales, the downscaling techniques and bias correction methods and the choice of future climate scenarios. Several different approaches were chosen for providing operational solutions to these drawbacks (Xu et al., 2005). However, numerous GCM simulations show almost univocal trends for global climate evolution. Giorgi & Lionello (2008) highlight a robust and consistent description specifically for the Mediterranean area, with a significant reduction in precipitation, mainly in summertime. In the same area, according to Giorgi (2006), a major increase in climatic variability is also expected.

Below, some results obtained by Senatore et al. (2011) are shown related to future water availability in the main basin of northern Calabria (Crati River Basin, 1332 km^2, Fig. 15) at the end of the XXI century. Future scenarios were made by applying the outputs of three Regional Climate Models (RCMs) RegCM, HIRHAM and COSMO-CLM to the newly developed Intermediate Space Time Resolution Hydrological Model (In-STRHyM). The analysis was performed using two time slices (1961–1990 and 2070–2099) with the SRES A2 (GCM HAD3AM) and A1B (GCM ECHAM5/MPI-OM) scenarios. Observed biases in simulated precipitation and temperature fields during the control period (1961-1990) were corrected before using meteorological outputs from each RCM as input for In-STRHyM.

In-STRHyM is a fully distributed hydrological model detailed enough to describe the hydrological processes of several small-medium sized Mediterranean basins. It has a relatively simple structure and is suitable for long period simulations to be undertaken within acceptable time frames. Specifically, In-STRHyM calculates separately transpiration and evaporation, depending on a remote sensing-derived vegetation fraction. Both transpiration and bare soil evaporation are estimated through the crop coefficient approach suggested by Allen et al. (1998), considering a water stress coefficient of the canopy depending on soil moisture conditions, and the reference values calculated through the Priestley & Taylor (1972) equation.

The RCMs predict an increase in mean annual temperature from 3.5 °C to 3.9 °C, and a decrease in mean annual precipitation from 9% to 21%. The effects of the changes in the forcing meteorological variables are relevant for all the hydrological output variables. Here we highlight results achieved for actual evapotranspiration (ET). This variable tends to decrease with reduced precipitation, but it increases with higher temperatures. Lower decrease in precipitation predicted by HIRHAM, together with the higher temperatures, leads to an average year ET increase of +2.5%, while for RegCM and CLM the annual mean reduction is equal to -5.1% (Fig. 15) and -8.3%, respectively. However, in the summer period, that is the irrigation period, in all cases an ET reduction is achieved (from -1.0% with HIRHAM to -9.1% with RegCM, Fig. 15), indicating a decrease in water availability for plants and soil. This water stress is better highlighted when considering simulated root zone soil moisture. For this variable a reduction is predicted, differently from ET, during the whole year (-20.7%±1.9%, -12.8%±1.9% and -17.6%±1.8% with RegCM, HIRHAM and CLM, respectively). Figure 16 shows as an example the daily changes computed using RegCM (the behavior considering the other RCMs is similar): they are less relevant in winter and spring, but the reduction is dramatic in summer and early autumn, due to the increased evaporative demand (up to -40% with RegCM).

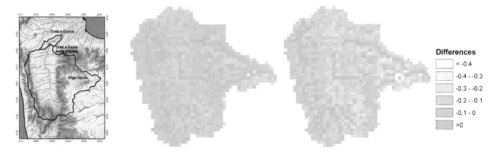

Fig. 15. Location of the Crati River Basin (left) and spatially distributed percentage changes in annual actual evapotranspiration (middle) and in actual evapotranspiration during the April–September irrigation period (right) simulated using RegCM (2070-2099 vs 1961-1990) (adapted from Senatore et al., 2011).

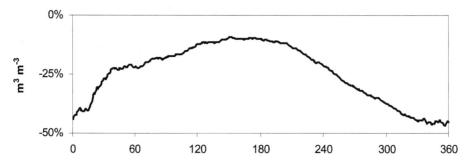

Fig. 16. Daily changes in root zone soil moisture computed using RegCM. RCM values are rescaled over 360 days, with the first day being October the 1st (readapted from Senatore et al., 2011).

7. Summary and conclusions

Evapotranspiration deeply affects the water resources availability in Calabria (average annual actual evapotranspiration estimated equal to almost 60% of the average cumulated annual rainfall). Highest water requirements come from agriculture, where losses due to evapotranspiration demand have to be re-equilibrated by huge amounts of water, mainly in the summer hot and dry period. The analysis of the comparison between the available water resource and the water demand was carried out considering both the "normal" conditions due to meteorological forcing, and the most critical derived by intense and prolonged drought periods. In the first case, neglecting the very conservative constraints proposed by the Regional Basin Authority for the minimum flow requirements, specific issues are not observed, the residual water availability being sufficient. Several problems arise instead when drought conditions occur: in these cases the development of guidelines is essential to define operative aspects about the individuation of the water use priorities, to characterize different drought levels, to individuate the main objectives of water management related to these levels, and to determine and apply the mitigation measures.

The proposed example of water resource management under shortage conditions in the agricultural area of the Sibari Plain shows the benefits that a proactive approach may provide with respect to the classical approaches based on emergency measures, which are usually expensive and not efficient. Within a proactive approach, specific care should be taken into account for reducing evapotranspiration losses through appropriate agronomic techniques. This action has to be considered as a strategic measure, with an impact on water scarcity reduction comparable to the effect of structural measures.

The review of drought indices showed that evapotranspiration could provide useful insights: i) when adopted within comprehensive indices, considering the effects of the whole water balance, and not only of some components, on water resources availability; ii) and mainly, when dealing with optical remote sensing techniques, because these allow to estimate in a relatively easy way the spatially distributed actual evapotranspiration over a specific area, and then they can relate this quantity to soil moisture and to the incoming of drought events.

Finally, applying some future scenarios with different GCMs and RCMs, it was observed that in Calabria the issues related to water resource management under shortage conditions in the next few years will be more frequent and intense, affecting wider areas. Evapotranspiration will be "tied down" by reduced precipitation (reducing its magnitude) and by higher temperatures (providing an opposite effect). It will not clearly increase or decrease on an annual basis, but in any case it will contribute to reduce useable water from the soil, needed for agricultural purposes. The hypothesized scenarios of climate change, though subject to uncertainty, have to be intended as an important part of knowledge for the planning of future interventions on the water resource by the Public Authorities, and for defining the optimal criteria to evaluate the amount of public investments.

8. References

Allen, R., Pereira, L.S., Raes, D. & Smith, M. (1998). Crop Evapotranspiration - Guidelines for Computing Crop Water Requirements. *FAO Irrigation and Drainage Paper* 56, FAO, Rome.

Allen, R.G., Tasumi, M. & Trezza, R. (2007). Satellite-based Energy Balance for MApping Evapotranspiration with Internalized Calibration (METRIC) – *Model. J. Irrig. and Drain. Engrg*, ASCE 133(4), 380-394.

Anderson, M., Norman, J., Diak, G., Kustas, W. & Mecikalski, J. (1997). A two-source time-integrated model for estimating surface fluxes from thermal infrared satellite observations. *Remote Sensing of Environment* 60: 195–216.

Anderson, M.C., Norman, J.M., Mecikalski, J.R., Otkin, J.A. & Kustas, W.P. (2007). A climatological study of evapotranspiration and moisture stress across the continental United States based on thermal remote sensing: 2. Surface moisture climatology. *J. Geophysical Research*; 112: art. No. D11112.

Bastiaanssen, W.G.M., Menenti, M., Feddes, R.A. & Holtslag, A.A.M. (1998) A remote sensing surface energy balance algorithm for land (SEBAL). 1. Formulation. *Journal of Hydrology* 212-213: 198–212.

Bastiaanssen, W.G.M. (2000). SEBAL-based sensible and latent heat fluxes in the irrigated Gediz Basin, Turkey. *Journal of Hydrology* 229: 87-100.

Bayarjargal, Y., Karnieli, A., Bayasgalan, M., Khudulmur, S., Gandush, C. & Tucker, C.J. (2006). A comparative study of NOAA–AVHRR derived drought indices using change vector analysis. *Remote Sensing of Environment*; 105: 9-22.

Bordi, I., Fraedrich, K., Petitta, M. & Sutera, A. (2005). Methods for predicting drought occurrences. *Proc. of the 6th EWRA International Conference*, Menton, France, 7-10 September 2005.

Cancelliere, A., Rossi, G. & Ancarani, A. (1996). Use of Palmer Index as drought indicator in Mediterranean regions. *Proc. IAHR Congress "From flood to drought"*, Sun City, South Africa, August 5-7, 1996, pp. S4.12. 1-25.

Cancelliere, A., Di Mauro, G., Bonaccorso, B. & Rossi, G. (2007). Stochastic forecasting of drought indices. In: Rossi, G., Vega, T., Bonaccorso, B. (Eds.), Methods and tools for drought analysis and management, *Water Science and Technology Library*, vol. 62, Springer, Dordrecht, 83-100.

COM – Commission of the European Communities, (2007). Addressing the challenge of water scarcity and droughts in the European Union, Communication from the Commission to the European Parliament and the Council, 414, 18.7.2007, Brussels, 14 pp.

COM – Commission of the European Communities, (2009). Towards a comprehensive climate change agreement in Copenhagen, Communication from the Commission to the European Parliament, the Council, the European Economic and Social Committee and the Committee of the Regions, 39, 28.1.2009, Brussels, 14 pp.

Cook, E.R., Seager, R., Cane, M.A. & Stahle, D.W. (2007). North American drought: Reconstructions, causes, and consequences. *Earth-Science Review*; 81: 93-134.

Cowie, G., Davis, M., Holmbeck-Pelham, S., Freeman, B., Freeman, M., Hatcher, K., Jackson, R., Miller Keyes, A., Merrill, M., Meyer, J., Sutherland, E. & Wenger, S. (2002). Reservoirs in Georgia: *Meeting Water Supply Needs While Minimizing Impacts*, University of Georgia's River Basin Science and Policy Center, Athens, Georgia,

Dworak, T., Berglund, M., Laaser, C., Strosser, Roussard, P. J., Grandmougin, B., Kossida, M., Kyriazopoulou, I., Berbel, J., Kolberg, S., Rodrigues-Diaz, J.A. & Montesinos, P. (2007). EU water saving potential, *Ecologic-Institute for International and European Environmental Policy*, ENV.D.2/ETU/2007/0001r.

Georgia Department of Natural Resources: Georgia Drought Management Plan, Atlanta, GA, (2003).

Gibbs, W.J. & Maher, J.V. (1967). Rainfall deciles as drought indicators. *Bureau of Meteorology Bulletin No. 48*, Commonwealth of Australia, Melbourne.

Giorgi, F. (2006). Climate change hot-spots. *Geophys. Res. Lett.* 33 L08707.

Giorgi, F. & Lionello, P. (2008). Climate change projections for the Mediterranean region. *Glob. Planet. Change* 63 90-104.

Guttman, N.B. (1998). Comparing the Palmer Drought Index and the standardized precipitation index. *J. Am.Water Res. Assoc.*; 34 (1): 113–121.

Hayes, M.J., Svoboda, M.D., Wilhite, D.A. & Vanyarkho, O.V. (1999). Monitoring the 1996 drought using the standardized precipitation index. *Bulletin of the American Meteorological Society*; 80: 429–438.

IPCC (2007). *Climate Change 2007. The Physical Science Basis.* Contribution of Working Group I to the Fourth Assessment Report of the Intergovernmental Panel on Climate Change [Solomon, S., D. Qin, M. Manning, Z. Chen, M. Marquis, K.B. Averyt, M. Tignor and H.L. Miller (eds.)]. Cambridge University Press, Cambridge, United Kingdom and New York, NY, USA, 996 pp., 2007.

Jackson, R.D., Idso, S.B., Reginato, R.J. & Printer, P.J. (1981). Canopy temperature as a crop water stress indicator. *Water Res.* 17, 1133-1138.

Karl, T.R. (1986). The Sensitivity of the Palmer Drought Severity Index (PDSI) and Palmer's Z-index to their calibration coefficients including potential evapotranspiration. *Journal of Climate and Applied Meteorology;* 25: 77-86.

Keyantash, J.A. & Dracup, J.A. (2004). An aggregate drought index: Assessing drought severity based on fluctuations in the hydrologic cycle and surface water storage. *Water Resources Research;* 40(9): Art. No. W09304.

Kim, T.W., Valdés, J.B., Nijssen, B. & Roncayolo, D. (2006). Quantification of linkages between large-scale climatic patterns and precipitation in the Colorado River Basin. *J. Hydrology;* 321: 173-186.

Kunstmann, H., Schneider, K., Forkel, R. & Knoche, R. (2004). Impact analysis of climate change for an alpine catchment using high resolution dynamic downscaling of ECHAM4 time slices, *Hydrology and Earth System Sciences,* Vol. 8, 1031-1045.

Lohani, V.K., Loganathan, G.V. & Mostaghimi, S. (1998). Long-term analysis and short-term forecasting of dry spells by Palmer Drought Severity Index. *Nord. Hydrol.;* 29(1): 21-40.

Matera, A., Fontana, G., Marletto, V., Zinoni, F., Botarelli, L. & Tomei, F. (2007). Use of a new agricultural drought index within a regional drought observatory. In: Rossi, G., Vega, T., Bonaccorso, B. (Eds.), Methods and tools for drought analysis and management, *Water Science and Technology Library,* vol. 62, Springer, Dordrecht, 103-124.

McKee, T.B., Doesken, N.J. & Kleist, J. (1993). The relationship of drought frequency and duration to time scales, *Proceedings of the 8th Conference on Applied Climatology,* American Meteorological Society, Anaheim, CA, Boston, MA, pp. 179–184.

Mendicino, G., Senatore, A. & Versace, P. (2005). I deflussi minimi annuali, stagionali e di magra nei corsi d'acqua calabresi, *Acts of the 25th Corso di Aggiornamento in Tecniche per la Difesa dall'Inquinamento,* ed. G. Frega, BIOS, Cosenza, pp. 89-117.

Mendicino, G. & Versace, P. (2007). Integrated Drought Watch System: A Case Study in Southern Italy, *Water Resour. Manage.,* Vol. 21, 1409–1428.

Mendicino, G., Senatore, A. & Versace, P. (2008a). A Groundwater Resource Index (GRI) for drought monitoring and forecasting in a Mediterranean climate, *J. Hydrol.,* Vol. 357(3-4), pp. 282-302.

Mendicino, G., Senatore, A. & Versace, P. (2008b). "Water resources management in agriculture under drought and water shortage conditions: a case study in southern Italy", *European Water 23/24,* pp. 41-56.

Menenti, M. & Choudhury, B.J. (1993). Parameterization of land surface evapotranspiration using a location dependent potential evapotranspiration and surface temperature range. In: Bolle H.J. et al. (Eds.), *Exchange Processes at the Land Surface for a Range of Space and Time Scales*, IAHS Publication 212, 561–568.

Moore, I.D., Norton, T.W & Williams, J.E. (1993). Modelling environmental heterogeneity in forested landscapes, *J. Hydrol.*, Vol. 150, 717–747.

Munda, G. (1995). Multicriteria Evaluation in a Fuzzy Environment, Series: *Contributions to Economics*, Physica-Verlag, Heidelberg.

Nakícenovíc, N., Alcamo, J., Davis, G., de Vries, B., Fenhann, J., Gaffin, S., Gregory, K., Grübler, A., Jung, T.Y., Kram, T., Emilio la Rovere, E., Michaelis, L., Mori, S., Morita, T., Pepper, W., Pitcher, H., Price, L., Riahi, K., Roehrl, A., Rogner, H.-H., Sankovski, A., Schlesinger, M.E., Shukla, P.R., Smith, S., Swart, R.J., van Rooyen, S., Victor, N. & Dadi, Z. (2000). *Special Report on Emissions Scenarios*, Cambridge University Press, Cambridge, available on-line: http://www.grida.no/climate/ipcc/emission/.

Narasimhan, B. & Srinivasan, R. (2005). Development and evaluation of Soil Moisture Deficit Index (SMDI) and Evapotranspiration Deficit Index (ETDI) for agricultural drought monitoring. *Agric. For. Meteorol.*; 133: 69-88.

Niemeyer, S. (2008). New drought indices. Options méditerranéennes, SERIE A: *Séminaiers Méditerranéens*, N. 80, "Drought Management: Scientific and Technological Innovations", pp. 267-274.

Norman, J., Kustas, W. & Humes, K. (1995). A two-source approach for estimating soil and vegetation energy fluxes from observations of directional radiometric surface temperature. *Agricultural and Forest Meteorology* 77: 263–293.

Ntale, H.K. & Gan, T.H. (2003). Drought indices and their application to East Africa. *Int. J. Climatol.*; 23: 1335–1357.

Palmer, W.C. (1965). Meteorological drought, Research Paper 45. U.S. Department of Commerce, Weather Bureau, Washington, DC.

Pereira ,L.S. (2007). Drought impacts in agriculture: water conservation and water saving practices and management, in: G. Rossi et al. (Eds.), *Water Science and Technology Library*, vol. 62: Methods and Tools for Drought Analysis and Management, Springer Netherlands, pp.349-373.

Priestley, C.H.B. & Taylor, R.J. (1972). On the assessment of the surface heat flux and evaporation using large-scale parameters, *Monthly Weather Review*, Vol. 100, 81–92.

Qian, T., Dai, A., Trenberth, K.E. & Oleson, K.W. (2006). Simulation of global land surface conditions from 1948-2004. Pt I: Forcing data and evaluations. *J. Hydrometeorol.*, 7, 953–975.

Rivas-Martinez, S. (1995). Bases para una nueva clasificacion bioclimatica de la Tierra. *Folia Botanica Matritensis 16.*

Roerink, G.J., Su, B. & Menenti, M. (2000). S-SEBI A simple remote sensing algorithm to estimate the surface energy balance. *Physics and Chemistry of the Earth (B)* 25(2): 147–157.

Rossi, G. (2003). An integrated approach to drought mitigation in Mediterranean regions, in: G. Rossi et al. (Eds.), *Tools for drought mitigation in Mediterranean regions*, Kluwer Academic Publishing, Dordrecht, pp. 3-18.

Rossi, G., Castiglione, L. & Bonaccorso, B. (2007). Guidelines for planning and implementing drought mitigation measures, in: G. Rossi et al. (Eds.), *Water Science and Technology Library*, Vol. 62: Methods and Tools for Drought Analysis and Management, Springer Netherlands, pp.325-347.

Senatore, A., Mendicino, G., Smiatek, G. & Kunstmann, H. (2011). Regional climate change projections and hydrological impact analysis for a Mediterranean basin in southern Italy. *Journal of Hydrology*, 399(1-2), 70-92.

Shafer, B.A. & Dezman, L.E. (1982). Development of a Surface Water Supply Index (SWSI) to assess the severity of drought conditions in snowpack runoff areas. *Proceedings of the Western Snow Conference*, Colorado State University, Fort Collins, Colorado, 164-175.

Steinemann, A.C. & Cavalcanti, L.F.N. (2006). Developing multiple indicators and triggers for drought plans. J. *water resources planning and management*, ASCE 132(3): 164-174.

Su, Z. (2002). The surface energy balance system (SEBS) for estimation of turbulent heat fluxes. *Hydrology and Earth System Sciences* 6(1): 85–99.

Su, Z., Yacob, A., Wen, J., Roerink, G., He, Y., Gao, B., Boogaard, H. & van Diepen, C. (2003). Assessing relative soil moisture with remote sensing data: Theory, experimental validation, and application to drought monitoring over the North China Plain. *Physics and Chemistry of the Earth (B)* 28(1-3): 89-101.

Thornthwaite, C.W. & Mather, J.R. (1955). The water balance, *Climatology*, Drexel Inst. Of Technology, Centeron, New Jersey.

Tsakiris, G., Loukas, A., Pangalou, D., Vangelis, H., Tigkas, D., Rossi, G. & Cancelliere, A. (2007a). Drought characterization, in: *Drought Management Guidelines Technical Annex*, Options méditerranéennes, Série B: Etudes et Recherches, Numéro 58.

Tsakiris, G., Pangalou, D. & Vangelis, H. (2007b). Regional drought assessment based on the Reconnaissance Drought Index (RDI). *Water Resources Management*; 21(5): 821-833.

Water Scarcity Drafting Group (2006). Water scarcity management in the context of WFD, Salzburg, June 1-2.

Wilby, R.L., Wedgbrow, C.S. & Fox, H.R. (2004). Seasonal predictability of the summer hydrometeorology of the River Thames, UK. *J. Hydrology*; 295: 1-16.

WHO/Unicef (2006). Protecting and Promoting Human Health, in: *The 2nd UN World Water Development Report*: "Water, a shared responsibility", available on line: http://www.unesco.org/water/wwap/wwdr/wwdr2/.

Xu, C., Widén, E. & Halldin, S. (2005). Modelling Hydrological Consequences of Climate Change-Progress and Challenges. *Advances in Atmospheric Sciences*, vol.22, no.6, 789-797.

Yao, Y., Liang, S., Qin, Q. & Wang, K., (2010). Monitoring Drought over the Conterminous United States Using MODIS and NCEP Reanalysis-2 Data. *Journal of Applied Meteorology and Climatology*, 49, 1665-1680.

Yevjevich, V. (1967). An objective approach to definitions and investigations of continental hydrologic droughts. *Hydrology Papers*, Colorado State University, Fort Collins, pp. 23.

Yevjevich, V., Hall, W.A. & Salas, J.D. (1978). Drought research needs, *Water Resources Publications*, Fort Collins, Colorado.

Yevjevich, V., Da Cunha, L. & Vlachos, E. (1983). Coping with droughts, *Water Resources Publications*, Littleton, Colorado.

Zhang, P., Anderson, B., Tan, B., Huang, D. & Myneni, R. (2005). Potential monitoring of crop production using a satellite-based Climate-Variability Impact Index. *Agric. For. Meteorol.*; 132: 344-358.

10

Evapotranspiration of Grasslands and Pastures in North-Eastern Part of Poland

Daniel Szejba
Warsaw University of Life Sciences – SGGW
Poland

1. Introduction

The problem of plant water requirements and supply is of great importance to agricultural water management. It is crucial to determine and provide the water amount required in a certain region to support the plants assimilation function. The quantity of water required on a specific farm can be determined by analyzing the water balance, where precipitation and evapotranspiration are basic elements. Evapotranspiration data is also indispensable when mathematically modelling the water balance. The values of evapotranspiration can be obtained from lysimeter measurements. However, this measurement is labour intensive and also requires special equipment; thus, it is not widely applied. To address this problem, a number of methods of evapotranspiration estimation based on physical and empirical equations are available, where the quantity of evapotranspiration depends on other measured factors. Penman (1948) developed a method for determination of the potential evapotranspiration as a product of the crop coefficient for a certain crop in a certain development stage and the reference evapotranspiration (Łabędzki et al., 1996). Open water surface evaporation is the reference evapotranspiration used in this method. Currently, the method most widely applied in Poland for evapotranspiration estimation is a method called the "French Modified Penman method", which is a version of FAO Modified Penman method (Doorenbos & Pruitt, 1977), with the net radiation flux calculated by Podogrodzki (Roguski et al., 1988). Name of "Modified Penman method" is using in further part of this text. On the other hand, the Food and Agriculture Organization (FAO) recommends the Penman-Monteith method for evapotranspiration estimation (Allen et al., 1998). The aforementioned methods require relevant crop coefficients to estimate the potential evapotranspiration. Although crop coefficients for grasslands and pastures applicable to the modified Penman are available for Polish conditions (Roguski et al., 1988; Brandyk et al., 1996; Szuniewicz & Chrzanowski, 1996), the problem occurs when the potential evapotranspiration has to be calculated according to the FAO standards which require the Penman-Monteith method to be used. Both the methods (Modified Penman and Penman-Monteith) require meteorological data including: air temperature, humidity, cloudiness or sunshine and wind speed. If one or more of the required inputs are not available, then applying any of the two methods is difficult, perhaps even impossible. In such cases, the Thornthwaite method, developed in 1931, can be a viable alternative (Byczkowski, 1979; Skaags, 1980; Newman, 1981; Pereira & Pruitt, 2004). The Thornthwaite method is commonly used in the USA. This method requires only two basic climatic inputs that determine the solar energy supply and are necessary to estimate the potential evapotranspiration: air temperature and day length.

There are two objectives of this chapter. The first objective is to determine the crop coefficient needed when estimating the potential evapotranspiration with the Penman-Monteith method. The second objective is a comparative analysis of the potential evapotranspiration estimates obtained from the Thornthwaite method and the crop coefficient approach with Penman-type formula as a reference evapotranspiration.

2. Reviewing the selected methods for evapotranspiration estimation: Modified Penman, Penman-Monteith and Thornthwaite

It can be assumed, that the amount of a farm plants evapotranspiration depends on such factors as atmosphere condition, plants development stage and soil moisture. The interdependence of these factors is complex and difficult to describe mathematically. This dependence can be expressed as a product of following functions:

$$ET = f_1(M) \cdot f_2(P) \cdot f_3(S) \tag{1}$$

where:
M - atmosphere factors,
P - plant factors,
S - soil moisture factors.

Groups of atmosphere factors can be formulated as a reference evapotranspiration (ET_0), which characterises meteorological conditions in the evapotranspiration process and describes evaporation ability in the atmosphere. This factor determines the intensity of evapotranspiration process in the case of unlimited access to a water source, that is deplete of soil water:

$$f_1(M) = ET_0 \tag{2}$$

$f_2(P)$ function describes the influence of plant parameters such as: plant species, development stage, mass of above ground and underground parts, leaf area index (LAI), growth dynamics, nutrients supply, yield and frequency of harvesting. A group of these parameters is expressed as a crop coefficient (k_c), which is empirically determined in independently by soil moisture conditions:

$$f_2(P) = k_c \tag{3}$$

$f_3(S)$ function describes the influence of soil moisture and the availability of soil water for plants (as a soil water potential) on evapotranspiration amount. With our knowledge of soil physics and plant physiology knowledge, it can be assumed that evapotranspiration during sufficient water supply does not depend or slightly depend on soil moisture (Łabędzki et al., 1996, as cited in: Kowalik, 1973; Salisbury & Ross, 1975; Feddes et al., 1978; Rewut, 1980; Olszta, 1981; Korohoda, 1985; Więckowski, 1985; Brandyk, 1990). Sufficient water supply does not limit evapotranspiration and plant yield is defined as a soil moisture range between optimum water content (when air content equals at least 8 - 10% in root zone) and refill point (pF 2.7 - 3.0). In other words, sufficient water supply means easily available water or readily available water (RAW). Evapotranspiration reductions has a place, when

RAW becomes consumed by plants. The deciding factor of evapotranspiration reduction amounts is the difference between actual soil moisture content and soil moisture content when the evapotranspiration process fades (wilting point). Thus, it can be showed in general (Łabędzki et al., 1996, as cited in: Olszta et al., 1990; Łabędzki & Kasperska, 1994; Łabędzki, 1995):

$$f_3(S) = k_s(\theta) \qquad (4)$$

where:
$k_s(\theta)$ – soil coefficient as a function of soil moisture.
Summarizing, equation (1) can be noted as below, where ETa is called actual evapotranspiration:

$$ETa = ET_0 \cdot k_c \cdot k_s \qquad (5)$$

In cases when sufficient water supply does not limiting evapotranspiration ($k_s = 1$), actual evapotranspiration (ETa) equals potential evapotranspiration (ETp):

$$ETp = ET_0 \cdot k_c \qquad (6)$$

The problem becomes how to determine a reference evapotranspiration and a crop coefficient.

2.1 The reference evapotranspiration computing by the Modified Penman method

Penman (1948) estimated the evaporation from an open water surface, and than used that as a reference evaporation. This method requires measured climatic data on temperature, humidity, solar radiation and wind speed. Analyzing a range of lysimeter data worldwide, Doorenbos and Pruitt (1977) proposed the FAO Modified Penman method. These authors adopted the same approach as Penman to estimate reference evapotranspiration. They replaced Penman's open water evaporation with evapotranspiration from a reference crop. The reference crop was defined as "an extended surface of an 8 to 15 cm tall green grass cover of uniform height, actively growing, completely shading the ground, and not short of water". The reference evapotranspiration according to Modified Penman method commonly applied in Poland was calculated by the following algorithm. This algorithm was developed according to following literature: Roguski et al. (1988); Feddes & Lenselink (1994), Kowalik (1995), Kędziora (1999), Woś (1995), Łabędzki et al. (1996), Łabędzki (1997), Feddes et al. (1997) and van Dam et al. (1997). The parameters are as follows:
φ - latitude of meteorological station [°],
J – day number [-],
T – daily average air temperature [°C],
RH - daily average relative humidity [%],
h_i - anemometer level above ground level [m],
v_{hi} – average wind speed on 10 m level [m s⁻¹],
c – average daily cloudiness in 11 degree scale,
n – duration of direct sunshine [h],
R_a - solar radiation at the external atmosphere border [W m⁻²],
α - albedo, in case of a crop equals to 0.23 [-],
γ - the psychrometric constant equals to 0.0655 [kPa K⁻¹],

λ - latent heat of vaporization equals to 2.45 [MJ kg^{-1}],
σ - Stefan – Boltzmann constant equals to 4.903*10^{-9} [MJ m^{-2} K^{-4} d^{-1}],
G_{sc} – solar constant equals to 0.082 [MJ m^{-2} min^{-1}].
Saturation vapour pressure (e_d) [kPa]:

$$e_d = 0.6108 \cdot \exp\left(\frac{17.27 \cdot T}{T + 237.3}\right) \tag{7}$$

Actual vapour pressure (e_a) [kPa]:

$$e_a = \frac{RH}{100} \cdot e_d \tag{8}$$

The slope of the vapour pressure curve (Δ) [kPa °C^{-1}]:

$$\Delta = \frac{4098 \cdot e_d}{(T + 237.3)^2} \tag{9}$$

Wind speed on 10 m level above ground level (v_{10}) [m s^{-1}]:

$$v_{10} = \frac{v_{hi}}{\left(\dfrac{h_i}{10}\right)^{1/7}} \tag{10}$$

Solar declinations (δ) [rad]:

$$\delta = 0.409 \cdot \sin\left(\frac{2\pi}{365} \cdot J - 1.39\right) \tag{11}$$

Relative distance to the Sun (d_r) [-]:

$$d_r = 1 + 0.033 \cdot \cos\left(\frac{2\pi}{365} \cdot J\right) \tag{12}$$

Time from sunrise to noon (w_s) [rad]:

$$w_s = a\cos\left(-\tan\varphi \cdot \tan\delta\right) \tag{13}$$

Possible sunshine (N) [h]:

$$N = \frac{24}{\pi} \cdot w_s \tag{14}$$

Solar radiation at the external atmosphere border (R_a) [W m^{-2}]:

$$R_a = \frac{24 \cdot 60}{\pi} \cdot G_{sc} \cdot d_r \cdot \left(w_s \cdot \sin\varphi \cdot \sin\delta + \cos\varphi \cdot \cos\delta \cdot \sin w_s\right) \tag{15}$$

Relation between real radiation to possible radiation – in case when sunshine value is not available there is calculated according to Angstöm criteria:

$$\frac{n}{N} = 1 - \frac{c}{10}$$ (16)

The net incoming short wave radiation flux (R_{ns}) [W m^{-2}]:

$$R_{ns} = R_a \cdot (1 - \alpha) \cdot \left(0.209 + 0.565 \cdot \frac{n}{N} \right)$$ (17)

The net outgoing long wave radiation flux (R_{nl}) [W m^{-2}]:

$$R_{nl} = \sigma \cdot (T + 273.2)^4 \cdot \left(0.56 - 0.08 \cdot \sqrt{10 \cdot e_a} \right) \cdot \left(0.1 + 0.9 \cdot \frac{n}{N} \right)$$ (18)

The net radiation flux (R_n) [W m^{-2}]:

$$R_n = R_{ns} - R_{nl}$$ (19)

The aerodynamic factor (E_a) [mm d^{-1}]:

$$E_a = 2.6 \cdot (e_d - e_a) \cdot (1 + 0.4 \cdot v_{10})$$ (20)

Modified Penman reference evapotranspiration (ET_{MP}) [mm d^{-1}]:

$$ET_{MP} = \frac{\Delta}{\gamma + \Delta} \cdot \frac{R_n}{\lambda} + \frac{\gamma}{\gamma + \Delta} \cdot E_a$$ (21)

2.2 The reference evapotranspiration computing by the Penman-Monteith method

Among scientists is unanimous the consensus is that the best method of evapotranspiration calculation is a method proposed and developed by John Monteith (1965). Monteith's derivation was built upon that of Penman (1948) in the now well-known combination equation (combination of an energy balance and an aerodynamic formula). The equation describes the evapotranspiration from a dry, extensive, horizontally uniform vegetated surface, which is optimally supplied with water. This equation is known as the Penman-Monteith equation and it is currently recommending by FAO. Potential and even actual evapotranspiration estimates are possible with the Penman-Monteith equation, through the introduction of canopy and air resistance to water vapour diffusion. Nevertheless, since accepted canopy and air resistance may not be available for many crops, a two-step approach is still recommended under field conditions. The first step is the calculation of the reference evapotranspiration as an evapotranspiration of a reference crop for some steady parameters and soil moisture conditions. In the second step the actual evapotranspiration is calculated using the root water uptake reduction due to water stress. The reference crop is defined as "a hypothetical crop which is grass, with a constant, uniform canopy 12 cm tall, constant canopy resistance equals to 70 s m^{-1}, constant albedo equals to 0.23, in conditions of active development and optimally supplied with water" (Łabędzki et al., 1996; Feddes et al., 1997; van Dam et al., 1997; Allen et al., 1998; Howell & Evett, 2004, as cited in: Monteith, 1965). The Penman-Monteith reference evapotranspiration recommended by FAO was calculated by a similar algorithm shown in point 2.1. The difference between the Modified Penman and Penman-Monteith methods bases on solar radiation and an aerodynamic formula calculation in general. Named factors were calculated according to following formulas shown below (Feddes & Lenselink, 1994).

The following parameters were used:

H- altitude of meteorological station over sea level [m],

T_{Kmin} – daily minimum air temperature [K],

T_{Kmax} – daily maximum air temperature [K],

v – average wind speed on 2 m level [m s^{-1}],

σ - Stefan – Boltzmann constant equals to $5.6745*10^{-8}$ [W m^{-2} K^{-4}],

Solar radiation at the external atmosphere border (R_a) [W m^{-2}]:

$$R_a = 435 \cdot d_r \cdot \left(w_s \cdot \sin\varphi \cdot \sin\delta + \cos\varphi \cdot \cos\delta \cdot \sin w_s \right) \tag{22}$$

Solar radiation (R_s) [W m^{-2}]:

$$R_s = R_a \cdot \left[0.25 + \left(0.5 \cdot \frac{n}{N} \right) \right] \tag{23}$$

The net incoming short wave radiation flux (R_{ns}) [W m^{-2}]:

$$R_{ns} = (1 - \alpha) \cdot R_s \tag{24}$$

The net outgoing long wave radiation flux (R_{nl}) [W m^{-2}]:

$$R_{nl} = \left(0.9 \cdot \frac{n}{N} + 0.1 \right) \cdot \left(0.34 - 0.139 \cdot \sqrt{e_a} \right) \cdot \sigma \cdot \frac{\left(T_{Kmax}^4 + T_{Kmin}^4 \right)}{2} \tag{25}$$

The radiation factor (R_n') [mm d^{-1}]:

$$R_n' = 86400 \cdot \frac{(R_{ns} - R_{nl})}{\lambda} \tag{26}$$

The atmospheric pressure [p_a] [kPa]:

$$p_a = 101.3 \cdot \frac{(T + 273.16 - 0.0065 \cdot H)}{T + 273.16} \tag{27}$$

The psychrometric constant (γ) [kPa °C]:

$$\gamma = 1615 \cdot \frac{p_a}{\lambda} \tag{28}$$

Modified psychrometric constant (γ') [kPa °C]:

$$\gamma' = (1 + 0.337 \cdot v) \cdot \gamma \tag{29}$$

The aerodynamic factor (E_a) [mm d^{-1}]:

$$E_a = \frac{900}{(T + 275)} \cdot v \cdot (e_d - e_a) \tag{30}$$

And finally Penman-Monteith reference evapotranspiration (ET_{P-M}) [mm d⁻¹]:

$$ET_{P-M} = \frac{\Delta}{\Delta + \gamma} \cdot R'_n + \frac{\gamma}{\Delta + \gamma} \cdot E_a \qquad (31)$$

2.3 Crop coefficient

Potential evapotranspiration is calculated by multiplying ET_o by k_c, a coefficient expressing the difference in evapotranspiration between the cropped and reference grass surface. The difference can be combined into a single coefficient, or it can be split into two factors describing separately the differences in evaporation and transpiration between both surfaces. The selection of the approach depends on the purpose of the calculation, the accuracy required, the climatic data available and the time step with which the calculations are executed (Allen et al., 1998). Due to the purpose of this chapter, only the single coefficient approach is taken under consideration. The single crop coefficient combined the effect of crop transpiration and soil evaporation. The crop coefficient expresses crop actual mass and development stage influence on the evapotranspiration value, in sufficient soil moisture content. It is dependant on crop type, development stage and yield. The generalized crop coefficient curve is shown in Figure 1. Shortly after the planting of annuals or shortly after the initiation of new leaves for perennials, the value for k_c is small, often less than 0.4. The k_c begins to increase from the initial k_c value, $k_{c\ ini}$, at the beginning of rapid plant development and reaches a maximum value, $k_{c\ mid}$, at the time of maximum or near maximum plant development. During the late season period, as leaves begin to age and senesce due to natural or cultural practices, the k_c begins to decrease until it reaches a lower value at the end of the growing period equal to $k_{c\ end}$ (Roguski et al., 1988; Allen et al., 1998).

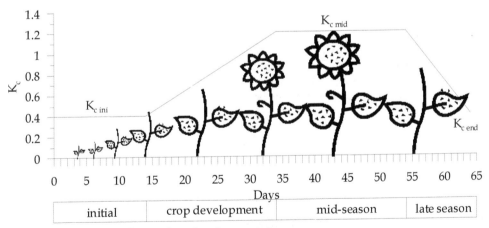

Fig. 1. Crop coefficient due to plant development stage

The objective of this work is to determine the crop coefficient needed when estimating the potential evapotranspiration with the Penman-Monteith method, when the potential evapotranspiration calculated as a product of Modified Penman reference evapotranspiration and appropriate crop coefficient for this method is known. Based on procedures proposed by Feddes et al. (1997), the conversion of the Modified Penman crop coefficient k_{cMP} to the Penman-Monteith crop coefficient k_{cP-M} can be write as:

$$ET_p = ET_{MP} \cdot k_{cMP} = ET_{P-M} \cdot k_{cP-M} \tag{32}$$

from which:

$$k_{cP-M} = \frac{ET_{MP} \cdot k_{cMP}}{ET_{P-M}} \tag{33}$$

2.4 Potential evapotranspiration estimation by the Thornthwaite method

Both Modified Penman and Penman-Monteith methods required many climatic inputs like: air temperature, relative humidity, wind speed and solar radiation or at least daily sunshine. These are limited or even not available for many regions. Another problem is noncontinuous data series for some periods. Thus using the Modified Penman and Penman-Monteith methods for evapotranspiration calculation is not so easy and problematic in some cases. An alternative commonly used in the United States is the Thornthwaite method, because it requires only air temperature as a input data (Skaags, 1980; Newman, 1981). This method is based on determination of available energy required for the evaporation process. The relationship between average monthly air temperature and potential evapotranspiration is calculated based on a standard 30 days month with 12 hours of daylight each day according to the following equation (Byczkowski, 1979; Newman, 1981; Pereira & Pruitt, 2004):

$$ETp_T = 16.2 \cdot \left(\frac{10 \cdot T_j}{I} \right)^a \tag{34}$$

where:
ETp_T – Thornthwaite monthly potential evapotranspiration (mm),
d_f – correction factor for daylight hours and days in month (-),
T_j – average monthly air temperature (°C),
I – annual heat index as a sum of monthly heat index I_i:

$$I = \sum_{i=1}^{12} I_i = \sum_{i=1}^{12} \left(\frac{T_j}{5} \right)^{1.514} \tag{35}$$

a – coefficient derived from climatological data:

$$a = 6.75 \cdot 10^{-7} \cdot I^3 - 7.71 \cdot 10^{-5} \cdot I^2 + 1.79 \cdot 10^{-2} \cdot I + 0.492 \tag{36}$$

In order to convert the estimates from a standard monthly ETp_T to a decade of evapotranspiration the following correction factor for daylight hours and days in month d_f (-) was used:

$$d_f = \frac{N_{dec}}{360} \tag{37}$$

where:
N_{dec} - possible sunshine for decade (h)
It must to be noted, that the Thornthwaite method is valid for average monthly air temperature from 0 to 26.5 °C.

3. Grasslands and pastures in the north-eastern part of Poland and local condition climate data

As Statistical Yearbook of Agriculture and Rural Areas (2009) presents, grasslands and pastures occupy about 3271.2 thousand hectares which is 20% of the total agricultural land in Poland. According to administrative division, the north-eastern part of Poland are Podlaskie and the eastern part of Warmińsko-Mazurskie voivodships. Grasslands and pastures occupy 393.5 thousand hectares (35%) and 290 thousand hectares (28.1%) of these voivodships agricultural land respectively. The valley of the River Biebrza, (22° 30'–23° 60' E and 53° 30'–53° 75' N) (Fig. 2) is one of the last extensive undrained valley mires in Central Europe. The Biebrza features several types of mires. The dominant types are fens, which account for some 75.9% of the wetland area (Okruszko, 1990). The altitude of the valley ranges from 100 to 130 m above mean sea level and the catchment area of approximately 7000 km² has a maximum altitude of 160 m (Byczkowski & Kicinski, 1984). The mean yearly rainfall is 583 mm, of which 244 mm falls in the wet summers. Mean annual temperature is rather low (6.8 °C), and the growing season is quite short (around 200 days) (Kossowska-Cezak, 1984). The part of Warmińsko-Mazurskie voivodship is Warmia region. Main town (former capital of Warmia region) situated on the north part of Warmia region (Fig. 2) is Lidzbark Warmiński (20° 35' E, 54° 08' N).

Fig. 2. An approximate location of considered regions in Poland

The altitude of the region ranges from 80 to 100 m above mean sea level on the borders and falls down from 40 to 50 m above mean sea level to the center. Brown Soils and Mollic Gleysols developed from silt and clay dominate in this. These soils are situated on sloping areas with partly well surface water outflow. In the study region average yearly air temperature is equal to 7.1°C and average yearly sum of precipitation equal to 624 mm. The highest amount of rainfall is usually observed in July and August. The vegetation period lasts about 200 days. The snow cover occurs during 60–65 days (Nowicka et al., 1994). The needed meteorological data are available for the 1989-2004 grassland growing seasons derived from the Biebrza meteorological station located in the Middle Biebrza River Basin. The estimation of the pasture evapotranspiration will be based on the meteorological data collected in the Warmia region during the 1999 through 2010 period.

4. Results and discussion

The decade Modified Penman and Penman-Monteith reference evapotranspiration values were calculated both for Warmia Region and Middle Biebrza River Basin. The relationship between reference evapotranspiration values of two kinds of Penman methods was shown on Fig. 3.

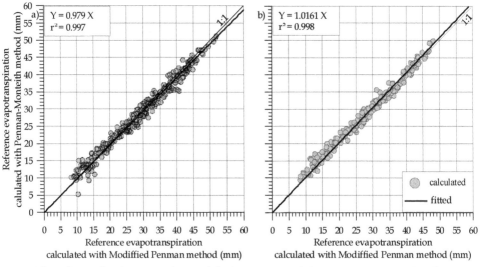

Fig. 3. The relationship between the Modified Penman and the Penman-Monteith reference evapotranspiration for: a) Middle Biebrza River Basin, b) Warmia Region

The relationship was fitted by linear regression through origin. Obtained linear equations indicates there is not significant difference between reference evapotranspiration calculated with Modified Penman and Penman-Monteith methods in both cases. It must to be noted that there is very good correlation between Modified Penman and Penman-Monteith methods. The coefficient of determination r^2 is equall to 99.7% and 99.8% respectively. Due to linear equation, Penman-Monteith reference evapotranspiration values are about 2% lower than values calculated by Modified Penman method for Middle Biebrza River Basin case (Fig. 3a). Whereas, an opposite situation was observed for Warmia Region. Reference evapotranspiration values calculated by the Modified Penman are 1.6% lower than values obtained by the Penman-Monteith method (Fig. 3b).

Consequently, an attempt was made for crop coefficient calculation (Eq. 33) proper for determination of potential evapotranspiration with the Penman-Monteith method. The following croplands were taken under consideration: pasture located in Warmia Region and intensive meadow, extensive meadow and natural wetland plant communities characteristic of Middle Biebrza River Basin. The calculation was conducted for vegetation period decade values of Modified Penman and Penman-Monteith reference evapotranspiration and crop coefficient for the Modified Penman method elaborated by Roguski et al. (1988), Brandyk et al. (1996) and Szuniewicz & Chrzanowski (1996). Considered values of crop coefficient both for Modified Penman ($k_{c\ MP}$) and Penman-Monteith ($k_{c\ P-M}$) for pasture was presented on Table 1. It can be maintain that $k_{c\ P-M}$ values for April are about 0.05 lower than $k_{c\ MP}$ values. The values for May, June and July are the same or almost the same – the difference does not exceed 0.02. The most significant differences are present in September, where $k_{c\ P-M}$ is lower than $k_{c\ MP}$ from 0.09 to 0.21.

Month	Decade	Crop coefficient	
		$k_{c\ MP}$	$k_{c\ P-M}$
April	1	0.75	0.70
	2	0.80	0.76
	3	0.80	0.76
May	1	0.85	0.84
	2	0.80	0.81
	3	0.95	0.95
June	1	0.70	0.71
	2	0.70	0.71
	3	0.95	0.97
July	1	0.80	0.81
	2	0.85	0.85
	3	0.90	0.89
August	1	0.80	0.79
	2	0.95	0.93
	3	1.05	1.00
September	1	0.95	0.86
	2	1.00	0.87
	3	1.10	0.89

Table 1. Crop coefficient of pasture for Modified Penman and Penman-Monteith methods

Modified Penman crop coefficient for extensive meadows (EM) and natural wetlands plant communities (NWPC) was published by Brandyk et al. (1996) as cited in: Roguski (1985) and Łabędzki & Kasperska (1994). Values of these crop coefficients as well as values of calculated Penman-Monteith crop coefficients was presented on Table 2. It can be maintain that $k_{c\ P-M}$ values are higher than $k_{c\ MP}$ values from 0.01 to 0.12 for extensive meadow in general. An exception to this rule is the last five decades, when $k_{c\,P-M}$ values are lower then $k_{c\,MP}$ values from 0.01 to 0.23. A similar tendency can be observed for natural wetland plant communities. But wider differences occur between $k_{c\,P-M}$ and $k_{c\,MP}$. A value of $k_{c\,P-M}$ is higher up to 0.08 than $k_{c\,MP}$ value for a few decades and lower until 0.31 for the last decade of September.

Month	Decade	Crop coefficient			
		EM		NWPC	
		$k_{c\,MP}$	$k_{c\,P-M}$	$k_{c\,MP}$	$k_{c\,P-M}$
April	1	0.93	1.05	0.62	0.70
	2	0.93	0.97	0.79	0.83
	3	0.85	0.84	0.75	0.74
May	1	0.88	0.90	0.77	0.79
	2	1.04	1.09	1.06	1.10
	3	1.03	1.08	1.21	1.27
June	1	0.76	0.79	1.24	1.30
	2	0.91	0.96	1.28	1.36
	3	0.98	1.04	1.40	1.48
July	1	0.99	1.03	1.32	1.37
	2	1.01	1.06	1.18	1.23
	3	0.98	1.04	1.40	1.48
August	1	0.97	0.98	1.30	1.31
	2	1.07	1.07	1.40	1.39
	3	1.18	1.15	1.40	1.36
September	1	1.34	1.27	1.63	1.55
	2	1.41	1.27	1.85	1.66
	3	1.41	1.18		1.60

Table 2. Crop coefficient of extensive meadow and natural wetland plant communities for Modified Penman and Penman-Monteith methods

The Modified Penman crop coefficient for intensive meadow located in Middle Biebrza River Basin was elaborated by Szuniewicz & Chrzanowski (1996). They based the research on lysimeter experiments conducted on peat –moorsh soil with a ground water level of 35 – 90 cm (optimum soil moisture) during the 1982-1991 period. Researchers had established conditions for 3-cut meadows with different hay yields: 0.10, 0.20, 0.30, 0.40 and 0.50 Mg ha^{-1}. The climate of the considered region is more severe compared to other plain regions in Poland, thus the vegetation period starts about two weeks later. Elaborated by Szuniewicz & Chrzanowski crop coefficients for the Modified Penman method as well as calculated crop coefficients for Penman-Monteith was presented on Table 2. There are not significant differences between $k_{c\,P-M}$ and $k_{c\,MP}$ values for the first two decades of the vegetation period.

The differences increase during successive decades of May and June from 0.02 up to 0.07. Next, they decrease from 0.04 to 0.02 in July. There are not significant differences again for first and second decades of July. The difference begins it's increase from the third decade of July up to the second decade of September. The values of $k_{c\ P-M}$ are even 0.12 – 0.18 lower than $k_{c\ MP}$ for the second decade of September. There is also a clear tendency towards an increase of differences between crop coefficients $k_{c\ P-M}$ and $k_{c\ MP}$ values due to an increase of potential hay yield. The $k_{c\ P-M}$ values get higher from 0.02 to 0.07 in May and June. However, the opposite tendency can be observed in September, when $k_{c\ P-M}$ get lower from 0.06 to even 0.18.

Month	Decade	Cut	Crop coefficient at hay yields Mg ha^{-1}									
			0.10		0.20		0.30		0.40		0.50	
			$k_{c\ MP}$	$k_{c\ P-M}$	$k_{c\ MP}$	$k_{c\ P-M}$	$k_{c\ MP}$	$k_{c\ P-M}$	$k_{c\ MP}$	$k_{c\ P-M}$	$k_{c\ MP}$	$k_{c\ P-M}$
April	2		0.93	0.96	0.93	0.96	0.93	0.96	0.93	0.96	0.93	0.96
	3		0.78	0.77	0.85	0.84	0.9	0.89	0.95	0.94	0.99	0.98
May	1	I	0.77	0.79	0.88	0.90	0.97	0.99	1.06	1.08	1.13	1.15
	2		0.89	0.93	1.04	1.09	1.17	1.22	1.28	1.34	1.39	1.45
	3		0.86	0.90	1.03	1.08	1.18	1.24	1.31	1.38	1.43	1.50
June	1		0.76	0.80	0.76	0.80	0.76	0.80	0.76	0.80	0.76	0.80
	2		0.86	0.91	0.91	0.96	0.95	1.01	0.99	1.05	1.02	1.08
	3	II	0.87	0.92	0.98	1.04	1.08	1.14	1.17	1.24	1.25	1.32
July	1		0.85	0.89	0.99	1.03	1.11	1.16	1.21	1.26	1.30	1.36
	2		0.86	0.90	1.01	1.06	1.15	1.20	1.27	1.33	1.38	1.44
	3		0.78	0.80	0.78	0.80	0.78	0.80	0.78	0.80		
August	1		0.89	0.90	0.97	0.98	1.04	1.05	1.09	1.10		
	2	III	0.95	0.94	1.07	1.06	1.17	1.16	1.26	1.25		
	3		0.96	0.94	1.18	1.15	1.36	1.33	1.52	1.48		
September	1		1.12	1.06	1.34	1.27	1.52	1.44	1.68	1.59		
	2		1.16	1.04	1.41	1.27	1.63	1.47	1.82	1.64		

Table 3. Crop coefficient of 3-cut meadow for Modified Penman and Penman-Monteith methods

The next step of this work use to be an comparison potential evapotranspiration calculated as a product of Penman-Monteith reference evapotranspiration and determined crop coefficient ($k_{c\ P-M}$) with alternative potential evapotranspiration by Thornthwaite. In order to solve the problem, decade values of Thornthwaite potential evapotranspiration was calculated (Eq. 34-37) and Penman-Monteith potential evapotranpiration applying crop coefficient for proper land use. The relationship between Thornthwaite potential evapotranspiration and Penman-Monteith potential evapotranspiration was presented on Fig. 4. The relationship was fitted by linear regression through origin. Analyzing obtained results, it can be maintain that Penman-Monteith evapotranspiration values are lower by about 25% for pasture (Fig. 4a) and 8% for extensive meadow than the Thornthwaite method

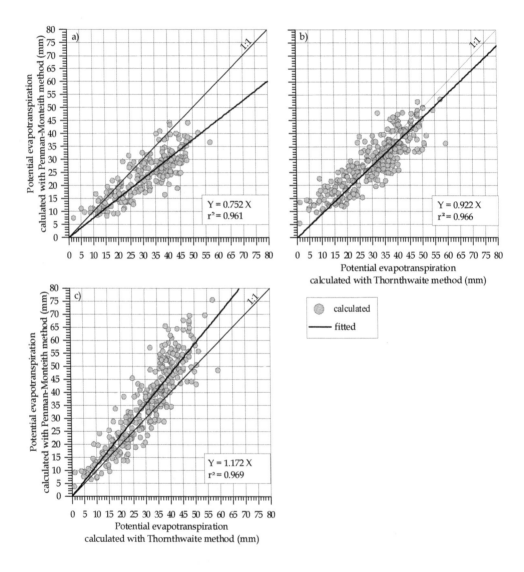

Fig. 4. The relationship between Thornthwaite potential evapotranspiration and Penman potential evapotranspiration for: pasture (a), extensive meadow (b) and natural wetland plant communities (c)

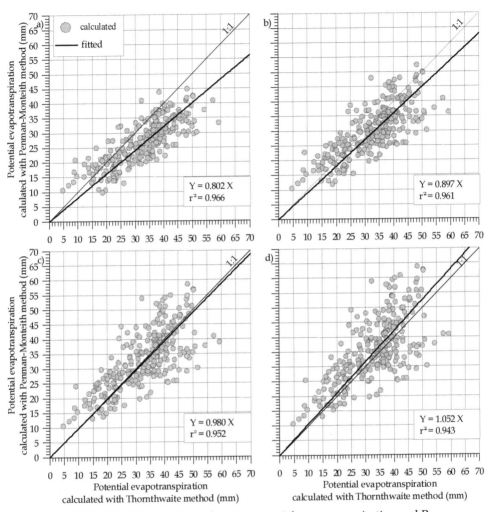

Fig. 5. The relationship between Thornthwaite potential evapotranspiration and Penman potential evapotranspiration of 3-cut meadow for hay yield Mg ha⁻¹: 0.10 (a), 0.20 (b), 0.30 (c) and 0.40 (d)

(Fig. 4b). Whereas in case of natural wetland plan community evapotranspiration, values calculated with Penman-Monteith method are of about 17% higher then values calculated with Thornthwaite method. It must to be noted, that coefficient of determination is almost equal ($r^2 \approx 97\%$) for all three cases. The relationship between Thornthwaite potential evapotranspiration and Penman-Monteith potential evapotranspiration for 3-cut meadow was presented on Fig. 5. Analyzing obtained results, it can be maintained that Penman-Monteith evapotranspiration values are very close to Thornthwaite evapotranspiration values for 0.30 Mg ha⁻¹ hay yield. An evapotranspiration calculated with the Thornthwaite method is just about 2% higher than Penman-Monteith evapotranspiration. The highest overestimation (20%) of the Thornthwaite method is observed for the lowest hay yield

(0.10 Mg ha^{-1}). The case of 0.20 Mg ha^{-1} hay yield characterizes about a 10% overestimation of the Thornthwaite method. An opposite case is the case of 0.40 Mg ha^{-1} hay yield, where the Thornthwaite method underestimates evapotranspiration by about 5%. Coefficients of determination vary between 94.3% (0.40 Mg ha^{-1} hay yield) and 96.6% (0.10 Mg ha^{-1} hay yield).

5. Conclusion

Based on the performed research the following conclusions can be formulated:
There are not significant differences between reference evapotranspiration calculated with the Modified Penman and Penman-Monteith methods of the Warmia Region as well as Middle Biebrza River Basin for entire vegetation period (April – September). Due to linear equation, Penman-Monteith reference evapotranspiration values are about 1.6 % higher than values calculated by the Modified Penman method for the Warmia Region case. Whereas, values of Modified Penman reference evapotranspiration are about 2.0% lower than values obtained with the Penman-Monteith method. From a practical point of view, the difference of total vegetation period reference evapotranspiration equals about 8 mm for the Warmia Region and 10 mm for Middle Biebrza River Basin due to 513 mm (Warmia Region) and 486 mm (Middle Biebrza River Basin) of average vegetation period reference evapotranspiration assumption.

Crop coefficients calculated for the Penman-Monteith evapotranspiration method are comparable or lower than crop coefficients for the Modified Penman method in case of pasture. Taking under consideration crop coefficient differences for extensive meadow and natural wetland plant communities it can be found that $k_{c\ P-M}$ values are higher than $k_{c\ MP}$ values from 0.01 to 0.12 for most of the vegetation period in general. An exception to this rule is the last five decades, when $k_{c\ P-M}$ values were lower then $k_{c\ MP}$ values from 0.01 even to 0.31. There are not significant differences between $k_{c\ P-M}$ and $k_{c\ MP}$ values for the first and second decades of vegetation period as well as for the first and second decades of July in the case of 3-cut meadow. The difference begins to from the third decade of July up to the second decade of September. The values of $k_{c\ P-M}$ are even 0.12 – 0.18 lower than $k_{c\ MP}$ for the second decade of September. Summarizing, crop coefficients calculated for Penman-Monteith method are almost equal or slightly higher compare to Modified Penman crop coefficients for most of a vegetation period in all considered land use. An exception are last three to four decades of vegetation period when values of $k_{c\ P-M}$ are clearly lower compared to $k_{c\ MP}$ values. These differences are equal during the entire vegetation period. But they can have essential meaning in certain parts (decades) of vegetation period when a crop water requirement is determined.

Potential evapotranspiration values calculated with the Thornthwaite method are overestimated in ratio to values calculated with the Penman-Monteith method in the following cases by about: 25% for pasture, 20% for 3-cut meadow (0.10 Mg ha^{-1} hay yield), 10% for 3-cut meadow (0.20 Mg ha^{-1} hay yield) and 8% for extensive meadow. Whereas, one time Thornthwaite potential evapotranspiration values were lower by about 5% for 3-cut meadow (0.40 Mg ha^{-1} hay yield). The best convergence of the considered methods is observed for 3-cut meadow in case of 0.30 Mg ha^{-1}. It has to be said, that coefficient of determination r^2 exceeds 94% of the value for all cases. Summarized, the Thornthwaite potential evapotranspiration method is comparable with the Penman-Monteith method for 3-cut meadow with a high value of hay yield and extensive meadow.

Future research should be focused on trials to find correlations between Thornthwaite and Penman-Monteith potential evapotranspiration for individual months of vegetation period. Another aim could be crop coefficient calculation for the Penman-Monteith method for field crops like grains, potatoes or sugar beets.

6. Acknowledgment

A part of this work considered to evapotranspiration calculation of Warmia Region was supported by the grant of Polish Ministry of Science and Higher Education No N N305 039234.
Special thanks to friend of mine Dr Jan Szatyłowicz for help with Penman's methods evapotranspiration calculation for Middle Biebrza River Basin.

7. References

Allen R.G., Pereira L.S., Raes D. & Smith M. (1998). Crop evapotranspiration - Guidelines for computing crop water requirements. *FAO Irrigation and Drainage Paper*, No. 56, pp. 290, ISBN 92-5-104219-5, FAO, Rome, Retrieved from: http://www.fao.org/docrep/x0490e/x0490e00.htm#Contents

Brandyk T., Szuniewicz J., Szatyłowicz J. & Chrzanowski S. (1996). Potrzeby wodne roślinności obszarów hydrogenicznych. *Zesz. Probl. Post. Nauk Rol.* 432, pp. 91-104, ISSN 0084-5477

Byczkowski A. (1979). *Hydrologiczne podstawy projektów wodnomelioracyjnych. Przepływy charakterystyczne*, ISBN 83-09-00035-9, Wyd. PWRiL, Warszawa

Byczkowski, A. & Kicinski, T. (1984). Surface waters in the Biebrza drainage basin. Pol. Ecol.Stud. 10, pp. 271–299, ISSN 0324-8763

Doorenbos J. & Pruitt W.O. (1977). Guidelines for predicting crop water requirements. *Irrigation and Drainage Paper*, No. 24, pp. 290, ISBN 92-5-100279-7, FAO, Rome

Feddes R.A. & Lenselink K.J. (1994). Evapotranpiration, In: *Drainage Principles and Application*. Ritzema H.P., (Ed.), ILRI, Publication 16, Second Edition, ISBN 90 70754 3 39, Wageningen, The Netherlands

Feddes R.A., Koopmans R.W.R. & Van Dam J.C. (1997). *Agrohydrology*, Wageningen University, Department of Environmental Sciences, Sub-department Water Resources

Howell T.A. & Evett S.R., (2004). *The Penman-Monteith Method*, USDA-Agricultural Research Service Conservation & Production Research Laboratory, Bushland, Texas, USA, Retrieved from: www.cprl.ars.usda.gov

Kędziora A. (1999). *Podstawy Agrometeorologii*, ISBN: 8309016417, Wyd. PWRiL, Poznań

Kossowska-Cezak U. (1984). Climate of the Biebrza ice-marginal valley. Pol. Ecol. Stud.10, 253–270, ISSN 0324-8763

Kowalik P. (1995). *Obieg wody w ekosystemach lądowych*. ISSN 0867-7816, Monografia PAN. Zeszyt 3., Warszawa

Łabędzki L., Szajda J. & Szuniewicz J. (1996). Ewapotranspiracja upraw rolniczych – terminologia, definicje, metody obliczania. *Materiały Informacyjne IMUZ*, pp. 1 - 7, Wyd. IMUZ, Falenty

Łabędzki L. (1997). Potrzeby nawadniania użytków zielonych – uwarunkowania przyrodnicze i prognozowanie. Wydawnictwo IMUZ Falenty

Newman J.E. (1981). Weekly Water Use Estimates by Crops and Natural Vegetation in Indiana. *Station Bulletin* No. 344, pp. 1-2, Department of Agronomy, Agricultural Experimental Station Purdue University. West Lafayette, Indiana

Nowicka A., Banaszkiewicz B. & Grabowska K. (1994). The selected meteorological elements for Olsztyn region in 1951–1990 years with comparison to averages for 1881–1930 period. *Mat. Konf. XXV zjazd agrometeorologow*, Olsztyn-Mierki, Poland, 27-29.09.1994

Okruszko H. (1990). *Wetlands of the Biebrza Valley, their Value and Future Management*, ISBN 83-00-03461, Polish Academy of Science, Warszawa

Pereira A.R. & Pruitt W.O. (2004). Adaptation of the Thornthwaite scheme for estimating daily reference evapotranspiration. *Agric. Water Manag.* 66, pp. 251-257, ISSN: 0378-3774

Roguski W., Sarnacka S. & Drupka S. (1988). Instrukcja wyznaczania potrzeb i niedoborów wodnych roślin uprawnych i użytków zielonych. *Materiały Instruktażowe* 66, pp. 90., ISSN 0860-0813, Wyd. IMUZ, Falenty

Skaggs R.W., (1980). Drainmod Reference Report. U.S. Department of Agriculture, Soil Conservation Service, North Carolina State University. Raleigh, North Carolina, pp. 19–23

Statistical Yearbook of Agriculture and Rural Areas (2009). ISSN 1895-121X, Zakład Wydawnictw Statystycznych, Warszawa

Szuniewicz J. & Chrzanowski S. (1996). Współczynniki roślinne do obliczania ewapotranspiracji łąki trzykośnej na glebie torfowo-murszowej w Polsce północno-wschodniej. *Wiad. IMUZ XVIII(4)*, pp. 109-118, ISBN 83-85735-28-3

Van Dam J.C., Huygen J., Wesseling J.G., Feddes R.A., Kabat P., Van Walsum P.E.V., Groenendijk P. & Van Diepen C.A. (1997). *Theory of SWAP version 2.0*, ISSN 0928-0944, Technical Document 45 DLO Winand Staring Centre, Wageningen

Woś A. (1995). *ABC meteorologii*. ISBN 8323207097, U.A.M. Poznań

Permissions

The contributors of this book come from diverse backgrounds, making this book a truly international effort. This book will bring forth new frontiers with its revolutionizing research information and detailed analysis of the nascent developments around the world.

We would like to thank Dr. Ayse Irmak, for lending her expertise to make the book truly unique. She has played a crucial role in the development of this book. Without her invaluable contribution this book wouldn't have been possible. She has made vital efforts to compile up to date information on the varied aspects of this subject to make this book a valuable addition to the collection of many professionals and students.

This book was conceptualized with the vision of imparting up-to-date information and advanced data in this field. To ensure the same, a matchless editorial board was set up. Every individual on the board went through rigorous rounds of assessment to prove their worth. After which they invested a large part of their time researching and compiling the most relevant data for our readers. Conferences and sessions were held from time to time between the editorial board and the contributing authors to present the data in the most comprehensible form. The editorial team has worked tirelessly to provide valuable and valid information to help people across the globe.

Every chapter published in this book has been scrutinized by our experts. Their significance has been extensively debated. The topics covered herein carry significant findings which will fuel the growth of the discipline. They may even be implemented as practical applications or may be referred to as a beginning point for another development. Chapters in this book were first published by InTech; hereby published with permission under the Creative Commons Attribution License or equivalent.

The editorial board has been involved in producing this book since its inception. They have spent rigorous hours researching and exploring the diverse topics which have resulted in the successful publishing of this book. They have passed on their knowledge of decades through this book. To expedite this challenging task, the publisher supported the team at every step. A small team of assistant editors was also appointed to further simplify the editing procedure and attain best results for the readers.

Our editorial team has been hand-picked from every corner of the world. Their multi-ethnicity adds dynamic inputs to the discussions which result in innovative outcomes. These outcomes are then further discussed with the researchers and contributors who give their valuable feedback and opinion regarding the same. The feedback is then collaborated with the researches and they are edited in a comprehensive manner to aid the understanding of the subject.

Apart from the editorial board, the designing team has also invested a significant amount of their time in understanding the subject and creating the most relevant covers. They scrutinized every image to scout for the most suitable representation of the subject and create an appropriate cover for the book.

The publishing team has been involved in this book since its early stages. They were actively engaged in every process, be it collecting the data, connecting with the contributors or procuring relevant information. The team has been an ardent support to the editorial, designing and production team. Their endless efforts to recruit the best for this project, has resulted in the accomplishment of this book. They are a veteran in the field of academics and their pool of knowledge is as vast as their experience in printing. Their expertise and guidance has proved useful at every step. Their uncompromising quality standards have made this book an exceptional effort. Their encouragement from time to time has been an inspiration for everyone.

The publisher and the editorial board hope that this book will prove to be a valuable piece of knowledge for researchers, students, practitioners and scholars across the globe.

List of Contributors

Carlos Krepper
Centro de Estudios Hidro-Ambientales-Facultad de Ingeniería y Ciencias Hídricas, Universidad Nacional del Litoral, Argentina
Consejo Nacional de Investigaciones Científicas y Técnicas, Argentina

Virginia Venturini and Leticia Rodriguez
Centro de Estudios Hidro-Ambientales-Facultad de Ingeniería y Ciencias Hídricas, Universidad Nacional del Litoral, Argentina

Ketema Tilahun Zeleke and Leonard John Wade
School of Agricultural and Wine Sciences, EH Graham Centre for Agricultural Innovation, Charles Sturt University, Australia

José Carlos Mendonça
Laboratório de Meteorologia (LAMET/UENF). Rod. Amaral Peixoto, Av. Brennand s/n Imboassica, Macaé, RJ, Brazil

Elias Fernandes de Sousa
Laboratorio de Engenharia Agrícola (LEAG/UENF); Avenida Alberto Lamego, CCTA, sl 209, Parque Califórnia, Campos dos Goytacazes, RJ, Brazil

Romísio Geraldo Bouhid André
Instituto Nacional de Meteorologia (INMET/MAPA); Eixo Monumental, Via S1 – Sudoeste, Brasília, DF, Brazil

Bernardo Barbosa da Silva
Departamento de Ciências Atmosféricas (DCA/UFCG); Avenida Aprígio Veloso, Bodocongó, Campina Grande, PB, Brazil

Nelson de Jesus Ferreira
Centro de Previsão de Tempo e Estudos Climáticos (CPTEC/INPE); Av. dos Astronautas, Jardim da Granja, São José dos Campos, SP, Brazil

Shakib Shahidian, Ricardo Serralheiro, João Serrano and Francisco Santos
University of Évora/ICAAM, Portugal

José Teixeira
Instituto Superior de Agronomia, Portugal

Naim Haie
Universidade do Minho, Portugal

Boris Faybishenko
Lawrence Berkeley National Laboratory, Berkeley, CA, USA

Tadanobu Nakayama
National Institute for Environmental Studies (NIES) 16-2 Onogawa, Tsukuba, Ibaraki, Japan
Centre for Ecology & Hydrology (CEH) Crowmarsh Gifford, Wallingford, Oxfordshire,
United Kingdom

Nirjhar Shah
AMEC Inc. Lakeland, FL, USA

Mark Ross and Ken Trout
Univ. of South Florida, Tampa, FL, USA

Zoubeida Kebaili Bargaoui
Tunis El Manar University, Tunisia

Giuseppe Mendicino and Alfonso Senatore
Department of Soil Conservation, University of Calabria, Arcavacata di Rende (CS), Italy

Daniel Szejba
Warsaw University of Life Sciences – SGGW, Poland

Printed in the USA
CPSIA information can be obtained
at www.ICGtesting.com
JSHW011423221024
72173JS00004B/651